GOLEMS AMONG US

Byron L. Sherwin

GOLEMS AMONG US

HOW A JEWISH LEGEND CAN HELP US NAVIGATE THE BIOTECH CENTURY

Ivan R. Dee

CHICAGO 2004

Library of Congress Cataloging-in-Publication Data:

Sherwin, Byron L.
    Golems among us : how a Jewish legend can help us navigate the biotech century / Byron L. Sherwin.
        p. cm.
    Includes bibliographical references and index.
    ISBN 1-56663-568-3 (alk. paper)
        1. Biotechnology—Moral and ethical aspects. 2. Genetic engineering—Moral and ethical aspects. 3. Golem. I. Title.

TP248.23.S535 2004
660.6—dc22

                                        2003068777

In memory of my father,

who taught me to question answers.

Contents

	Grandmother's Tales	3
1	Golems 'R Us	7
2	The Golem of Prague	18
3	Creating a Golem	27
4	Golems Among Us	35
5	The Past Meets the Present: The Golem and the Genome	47
6	Organic Golems: Frankenfood and Designer Genes	66
7	Test-tube Golems: Stem Cells and the Cloned Arranger	94
8	Mechanical Golems: Toward a Postbiological Human Future?	123
9	Corporate Golems: The Supreme Court Creates an Artificial Person	156
10	The Honey and the Sting: The Golem Meets Frankenstein	192
	A Note on Sources and a Personal Note	213
	Index	221

"Der oilem iz a goilem."

(The world is a golem.)

—*Anonymous Yiddish proverb*

GOLEMS AMONG US

Grandmother's Tales

"What are your earliest memories?" The great psychoanalyst Alfred Adler would pose this question to his patients. Adler recognized that childhood memories play a decisive role in shaping our lives.

Among my earliest memories are the taste of homemade sponge cake my grandmother would feed me before I went to bed, the Yiddish and Polish lullabies she would sing to me as she tucked me in, and the stories she would tell me before I fell asleep. Often, when fast asleep in my grandparents' small apartment in the Bronx, I would relive these stories in my dreams.

One night when I was four or five years old, Grandma told me the story of the golem. More than half a century has passed since then, and still I dream about the golem. Through the years, in my dreams and while awake, I have searched for the golem. This book is an invitation to join me in that quest.

At home my American-born-and-bred parents would read me stories of King Arthur and his knights of the Round Table, of

Robin Hood and his merry men, of George Washington and his cherry tree, of Snow White and the dwarfs. The stories my parents read to me were about places and people that seemed distant and detached, places they had never visited, people they had never known, and events that were figments of the imagination of the authors whose books they held in their hands. But Grandma's stories were not read; they were told. Her stories seemed to recount tales of a realm she had just left, places she just had been, people she knew well. Grandma told her tales like a native guide introducing a tourist to his native land. Her stories flowed from childhood memories of what and whom she had seen and heard in a far-off land: pre–World War I Poland.

The stories Grandma conveyed were tales of a lost Jewish Atlantis, a Jewish kingdom in a distant land that once was but was now no more. In the years just before my birth, a fire-spewing dragon had decimated its inhabitants. All that remained were people like Grandma, who had left before the flood of fire we now call the Holocaust had descended upon that realm. All that remained were Grandmother's memories, and Grandmother's tales.

In Yiddish the term *bube-meises* or "grandmother's tales" refers to stories not to be taken too seriously—"tall tales." But to the child I once was, and to the adult I have become, Grandma's stories became part of me. As Hasidism teaches, through hearing a tale as it is retold, words enter the body through the ear and become incorporated into one's self. Tales thereby become part of us. By retelling a tale, a person can become a participant in the story as it is told. In telling the tale, those within it live once more. Undoubtedly that is why Grandma would constantly remind me that many of the erudite and wonder-working rabbis who populated her stories were her ancestors, and mine as well.

Listening to these legends was an invitation to be part of a family reunion.

Grandma's stories were not meant simply to entertain me or to lull me off to sleep. Her stories were meant to be gifts, part of a precious legacy, family heirlooms, given to me to preserve and transmit. Grandma's tales were memories of fastly fading photos in an otherwise soon to be discarded family album.

No one I knew was a descendant of King Arthur or Robin Hood. No one I ever met had known Merlin or Snow White. Yet in Grandma's stories I met my forebears. Among them were scholars who had compiled erudite tomes that I would later find leaning on dusty library shelves anxiously awaiting a rare reader. Also among them were rabbis who had once been revered for the wisdom and clairvoyance they had displayed, for the exorcisms they had performed, and for the golems they had made.

On the night Grandma told me in graphic detail how her great-grandfather, the Rabbi of Makow, had exorcised a *dybbuk* from a woman who had been possessed, I could not sleep. I stayed awake, afraid that a *dybbuk* might invade my body while I slept. The following night I pleaded with her not to tell me any stories. But she insisted. That night she told me the legend of the golem. Knowing that the golem could protect me—even from a *dybbuk*—I promptly fell asleep. I dreamed about making a golem, about how the golem would accompany me wherever I might go. Who needed a fairy godmother if you had a golem?

In all the nights since that night, I have never met King Arthur or Snow White, but I have encountered many golems. Inevitably, so have you, though you may not have recognized them as such. In the pages that follow, I introduce some of them that are particularly relevant to our times.

According to an old Yiddish proverb, *Der oilem iz a goilem*—
The world is a golem. More than ever before, this observation is
true of our world, of our times. For reasons that will become apparent in what follows, our juncture in history might be accurately described as the age of the golem. No longer simply a
bube-maise, a grandmother's tale, the golem has emerged from
the shadows of legend to populate our world. Golems now dwell
among us; I no longer have to search to find them. They are all
around us in a wide variety of sizes and guises. Every day we coexist with them. How we learn to live with them will shape not
only their future but ours as well. As the early-twentieth-century
Austrian novelist Gustav Meyrink wrote in *The Golem*, his now-
classic surrealistic novel:

> The Golem? I've heard talk of it a lot.
> What do you know about the Golem? . . .
> Always they treat it as a legend,
> till something happens and turns it
> into actuality once more.

Before confronting the golems in our midst, before seeing
how the legend of the golem addresses problems and issues that
confront us today, it is first necessary to understand what the
golem legend is and how it developed from ancient times until our
own day. That is the task of the three chapters that now follow.

Chapter One

Golems 'R Us

According to Scripture, the first human be-
ing, Adam, was created by God on the sixth day of creation. But
according to rabbinic lore, Adam began his existence as a golem.

The order of events on the sixth day of creation is recounted
in various rabbinic texts. In each, an hour-by-hour description is
given of God's activities on that day. In some such texts, the first
hours of the sixth day are described as follows.

In the first hour, God conceives of creating human beings.
In the second hour, God takes counsel with the angels regarding
the advisability of the plan to create human beings. The Talmud
reports that the angels advised God against proceeding with the
plan: the risks, in their view, outweighed the benefits. The rebel-
lious and violent nature of human beings, their proclivity to sin
and moral vice, made their creation undesirable and unworthy of
God's creative activities. Despite their advice, God proceeds with
his plan.

In the third hour of the sixth day, God assembles virgin dirt from various places around the world. The first human being would not be linked to any one place or to any particular land. The earth he would be created from would be gathered from all over the world. In the fourth hour, God kneads the dirt. In the fifth, God shapes the form of the creature. In the sixth, God "makes it into a golem." In the seventh hour, God breaths a soul into the creature. At this point—with the infusion of a soul—the first golem becomes the first human being.

Adam is a golem who has become a human being by being granted a soul by God. Although a living being in human form, a golem lacks certain essential characteristics that would make it fully human. A golem is an incomplete human being; a human being is an evolved golem. According to these ancient rabbinic texts, a golem is the ultimate ancestor of the human race. Who, then, is a golem? Golems are us.

Ironically, though the etymological root of the Hebrew word *golem* refers to something unformed or amorphous, many of the rabbinic texts describing the creation of the first golem by God depict its physical form as approximating that of a human being. *Golem*, which denotes having no form, is nonetheless described as having a particular form, a human form. While the term *golem* does not appear in the Bible, a form of the word, *golmi*, is found in Psalms (139:16): "Your [that is, God's] eyes saw my unformed mass (*golmi*), it was all recorded in Your book."

In rabbinic Hebrew, *golem* takes on a number of meanings, denoting incompleteness. The biblical use of the root *glm*, to mean an unformed mass, is expanded to include both things (for example, vessels) and people that are somehow incomplete. A person who is underdeveloped socially, sexually, intellectually, or morally is described as a golem, an incomplete human being.

Similarly, in Yiddish slang a dull, boring, stupid, and humorless person is commonly referred to as a golem.

The description of a golem as a proto–human being, and the depiction of a golem as a somehow incomplete human being, raises three vital questions that permeate ancient, medieval, and modern reflections on the golem legend and that continue to confront us today:

1. What is the relationship—that is, the similarities and differences—between golems and human beings?

2. Could a humanly created golem achieve what the first divinely created golem was granted, human status and personhood? Can golems become humans?

3. Can humans devolve back to the golemic state from which they once emerged? Can humans become golems?

As we review the nature and the implications of the golem legend in later chapters, these three questions will often reemerge.

The Talmudic rabbis describe God as being involved in the creation not only of the first human being but also in the creation of all human beings who followed. According to the Talmud, there are three partners in the creation of each human being: father, mother, and God. Human parents provide the child with a body while the divine parent provides the child with a soul. In ancient and medieval Hebrew, *golem* often denotes either a human body or an embryo. Hence human parents create a child as a golem—an embryo—while God supplies the key element that makes the child a human being—the human soul. In this view, the birth of each human being reenacts the creation of the first human being in that God provides a golem with a human soul. But, unlike the story of the sixth day of creation, in the birth of each child God has human partners in the act of creation and procreation.

According to rabbinic literature, God created only one golem, the one brought into being on the sixth day of creation. Later on, however, golems are created by human beings. What is the nature and the potential of these humanly created golems? Much of the discussion of this question is found in commentaries on a Talmudic text that reports the creation of a golem by a human being.

The Talmudic text is terse and precise: "Rava said: If the righteous desired it, they could create worlds, for it is written, 'Your iniquities have distinguished between you and your God' (Isaiah 59:2). Rava created a man and sent him to Rabbi Zeira. Rabbi Zeira spoke to him [to the man], but received no answer. Thereupon, he [Rabbbi Zeira] said to him [to the man], 'You are from the companions. Return to your dust.'" And, presumably, the "man" did so.

Unlike most Talmudic texts that stimulate debate and discussion, this text is presented in a matter-of-fact manner. Surprisingly, the propriety of Rava's creation of the "man" is neither debated nor discussed. This particular text is part of a larger discussion of forbidden magical activities, such as sorcery and witchcraft, which are prohibited by biblical law. Yet Rava's creation of an "artificial man" is neither questioned, condemned, nor prohibited. The propriety of the human creation of life is presumed by Talmudic tradition.

Not so in Christian or European cultural tradition. It is told, for example, that Albertus Magnus created an anthropoid who acted as his servant. St. Thomas Aquinas denounced and destroyed this anthropoid as the work of the devil. Similarly, in the legend of Faust, it is the devil who gives Dr. Faust the power to create life. Later, in Mary Shelley's *Frankenstein*, Dr. Frankenstein describes his actions to create life as evil and odious, as dabbling in the "unhallowed arts." Contemporary opponents of various forms of genetic engineering and reproductive biotechnology denounce such activities as morally reprehensible, as unnatural, as "playing God."

Some equate "playing God" with the taking of unwarranted risks by engaging in activities with inevitably dangerous consequences. This is the view of *Frankenstein*. But it is not the Jewish view. Jewish tradition understood that creativity and innovation, by their very nature, entail risk but need not engender catastrophe. After all, the Talmudic rabbis raise no objection to Rava's creation of the "man." The Talmud also describes how God proceeded with the unprecedented creation of the first golem and the first human being with full awareness of the risks and consequences of that creative endeavor. God did so, according to the Talmudic legend, against the best advice of the angels. Perhaps the angels—guardians of the status quo—did not want God to "play God."

At the point in Genesis where the human being is described as having been created in God's image, all we know of God is that God creates worlds, plants, creatures, and human beings. The biblical claim that human beings have been created in the divine image is interpreted by the Talmudic rabbis to mean that the human being can share in the divine creative power to create life, including human life, and even new life-forms. Indeed, in the Jewish mystical tradition, human participation in such endeavors may induce a desirable state of spiritual rapture in which the human being can share with God the experience of ultimate creativity.

The creation of various life-forms by human beings is not considered by classical Jewish literature to be unnatural or ethically problematic. The great sixteenth-century sage, Rabbi

Judah Loew of Prague, who plays a central role in modern retellings of the golem legend, considered such activities as extensions of nature rather than as unnatural. According to Judah Loew, God created nature in an unfinished and incomplete form. The human task is to complete "the work of creation." Such activities are not considered human usurpations of divine prerogatives but rather as the work of human beings acting in their divinely mandated role as "God's partner in the work of creation." Because such deeds imitate divine creativity and nature, they cannot be considered unnatural. In Loew's words, "Human beings bring to fruition things not previously found in nature; nonetheless, since these are activities that occur in nature, it is as if they had entered the world to be created." In other words, God arrested the process of creation before its completion. The human task is to develop the potentials of the raw materials in nature created by God.

Although the Talmudic text noted above tells us that Rava created a "man," a golem, it does not tell us how he did so. That task is left to the medieval commentators. If creating a golem is considered to be imitative of God's creation both of the world and of the first golem, it should not be surprising that the medieval Jewish mystics incorporated various features of God's creative processes into the creation of a golem. Like the very first golem, most golems are created from dust. But the formation of a manikin from soil does not a golem make. It needs to be animated, to be transformed from inorganic to organic matter.

According to the biblical story of creation, God creates the world through speech. "God spoke and there was . . ." This indicated to the Jewish mystics that words, letters, speech, held the mystery of the power to create. Since the rabbis assumed that God spoke in Hebrew, the Jewish mystics believed the

Hebrew alphabet to be pregnant with creative power. By mastering certain particular combinations of Hebrew letters, creation could ensue, inert matter could be transformed into a living creature. The mystics believed that these secret formulas, the recipes for combining the letters to create life, could be found in early mystical treatises like the *Sefer Yetzirah*, The Book of Creation.

In the Talmudic account, Rava sends the "man" he had created to Rabbi Zeira. When Rabbi Zeira speaks to the man and receives no reply, Rabbi Zeira causes him to return to the dust from which he has been created. According to the medieval commentators, he does so through speech. Just as a golem is animated by the recitation of certain combinations of consonants and vowels, one way to deactivate a golem is to recite the same combinations backward. In other words, language has both the power to create as well as the power to destroy. But why did Rabbi Zeira destroy Rava's golem?

Rabbi Zeira seems to have considered Rava's creative endeavor to have been a failure. Rava tried to create a human being but failed, and managed to create only a golem. When Rabbi Zeira spoke and received no reply, he discerned that the "man" sent by Rava was not a man at all but a golem—a soulless and speechless anthropoid. As the medieval commentaries note, the story of Rava's creation is introduced by the verse from Isaiah (59:2), "Your iniquities have distinguished between you and your God." In other words, Rava's sins prevented him from creating a complete human being. The being created by Rava was flawed because Rava was flawed. Human creations thus reflect both the character and technical skill of their creator. Our fabrications are mirrors of our own best selves and our own worst selves. They embody our virtues as well as our vices.

If Rava tried to create a human being, he failed. Yet if he tried only to create a golem, he succeeded. According to classical Jewish sources, though Rava was the first human creator of a golem, he was not the last. In ancient and medieval versions of the golem legend, mystics did not create golems for practical reasons; that came later. Instead they created golems for no reason other than to achieve communion with God, sharing the experience of the creative moment when God had created life in a similar manner. In some medieval Jewish mystical texts, creation of a golem is considered part of an initiation rite into the ranks of the spiritually adept. Only in modern versions of the golem legend, as we shall see in the following chapter, are golems described as having been created for practical activities—to be a servant in the home or to protect innocent people from harm and violence. Unlike some recent versions of the golem legend, where golems are depicted as dangerous and demonic monsters, ancient and medieval versions describe the golem as a silent witness to the creativity that human beings can share with God. Ironically, though created with language, the golems of the early versions of the legend neither speak nor respond to speech. Once created, a golem has served its purpose simply by having been being created and thereby demonstrating the mystical and creative capabilities of its maker. It may then be destroyed. In this view, Rabbi Zeira would be held blameless for having destroyed Rava's golem. Yet for the Talmudic commentators, it was not so simple.

The commentators agreed that Rava's golem was flawed because Rava was morally flawed and perhaps also technically inept. But they debated the issue of whether someone more upright and skilled than Rava might have been able to create a golem that could become human in all respects. Two views emerged regarding the question of whether human beings can create other human beings, whether—as in the early rabbinic texts describing God's creation of the first golem who became human—a golem could evolve into a human being. Predictably, the two views are: yes and no. In classical Jewish literature this issue is brought to a head in discussions of an early eighteenth-century responsum (a decision in case law) by Rabbi Zevi Ashkenazi of Amsterdam, better known as Hakham Zevi.

Hakham Zevi raises the question of the legal status of a golem, specifically, can a golem be included in a *minyan*, a quorum for prayer? What Hakham Zevi is really asking is whether a golem can attain human status and, if not, what legal status might a golem have. Indeed, notes Hakham Zevi, if a golem can attain human status, Rabbi Zeira might have been guilty of murder for having destroyed Rava's golem. Hakham Zevi decides that because Rava's golem was not of woman born (in a different sense than Banquo in Shakespeare's *Macbeth*), it cannot be considered human and therefore cannot be counted in a quorum for prayer. Although Hakham Zevi does not grant actual or even potential human status to a golem, he does not tell us what kind of status a golem has. But, his son, Rabbi Jacob Emden, does. In one of his works, Emden describes a golem as "an animal in the form of a man." For both Ashkenazi and Emden, because a golem is not a human being, Rabbi Zeira is not guilty of murder. But both recognize that the golem is a living being and that Rabbi Zeira might have acted inappropriately in destroying it for no apparent reason.

In the late nineteenth century the Hasidic master Rabbi Gershon Hanokh Leiner of Radzyn took issue with Ashkenazi's and Emden's views that denied either actual or potential human status to a golem. Leiner claimed that if Rava had not been morally flawed, he could have created a golem that was not flawed—he could have created a golem that either was or could have become human in every respect. In addition, Leiner reminds his readers that Adam—whom God initially created as a golem—was not of woman born, and that Adam was undeniably human, so that killing him surely would have been murder. Unlike Ashkenazi, Leiner does not consider the method by which a creature is conceived and gestated to be the primary determinant of its legal status. For some commentators, like Leiner, a golem can attain human status. For others, like Emden, it cannot; it has the status of an animal. For still others, golems have a unique legal status of their own. The issue of what precisely is the legal status of a golem will become relevant in later chapters when we consider the nature of some of the golems that currently populate our world.

In his responsum, Zevi Ashkenazi refers to a golem created by his ancestor, the sixteenth-century Polish rabbi, Elijah of Helm. According to various versions of this story, Rabbi Elijah created a golem to be his servant. When the golem continued to grow in strength, power, and size, Rabbi Elijah realized that it posed a potential danger to life and property. He then destroyed the golem by removing a piece of parchment from its body on which divine names that had animated it had been written. According to versions of this story preserved in non-Jewish sources, the golem reverted back to the clay from which it had been formed, and collapsed on the rabbi, either severely injuring or killing him. According to the version preserved by his descendant, Rabbi Jacob Emden, however, the collapsing golem merely scratched Rabbi Elijah on the face.

This version of the golem legend initiated a transition from how the golem was understood in ancient and medieval Jewish literature to how the golem has been understood and portrayed in more recent times. In the story of Rabbi Elijah's golem, we encounter three motifs not found in earlier sources. The first is that the golem is created to serve a certain practical purpose, in this case to be a servant. The second is that the golem is considered to be potentially dangerous. The third is that the golem can harm its creator. The tale of Rabbi Elijah's golem introduces the theme, greatly amplified in modern literature, that the artifacts we create to help us may end up either harming or destroying us.

The story of the creation of a golem by Rabbi Elijah of Helm became the blueprint for the most famous and influential version of the golem legend, the creation of a golem by Rabbi Elijah's sixteenth-century contemporary, Rabbi Judah Loew of Prague. It is the story of Rabbi Loew's creation of the golem of Prague that became the prototype for most expressions of the golem legend in modern literature and art. It is to the legend of Judah Loew and the golem of Prague that our attention now turns.

Chapter Two

The Golem of Prague

Rabbi Judah Loew of Prague was a supernova in the bright constellation of sixteenth-century Jewish scholars and communal leaders. His mystical, theological, ethical, and legal writings are extensive and profound. Since his death in 1609, the legacy of Loew's life and thought has been preserved by a small cadre of Jewish scholars. His works are still studied today in *yeshivot* around the world. Yet the life, work, and thought of Judah Loew of Prague have been largely eclipsed by the legend of his creation of the golem of Prague. According to this legend, Judah Loew created the golem of Prague on March 20, 1580.

In none of Loew's voluminous writings is there any hint of his ever having created a golem. Although he commented on the Talmudic story of Rava's golem, Loew never seems to have considered creating one of his own. Neither in the writings of Loew's contemporaries nor in the writings about him for well over two hundred years after his death is there any reference to his having created a golem. Then, suddenly, in the mid-nineteenth century, stories began to emerge about the golem created by Rabbi Judah Loew of Prague. Subsequently, Judah Loew becomes the archetypal creator of the golem, and the golem he is alleged to have created becomes the prototype for golems described in generations yet to come.

Unlike the golems of ancient and medieval Jewish tradition, the golem of Prague became the model for most modern tellings of the golem legend. Unlike its predecessors described in classical Jewish literature, the golem of Prague is depicted as a servant, protector, warrior, and savior. As the modern version of the legend developed into the twentieth century, the golem emerged as a source of danger, a reckless creature, a Jewish facsimile of the monster created by Victor Frankenstein. Indeed, these more recent descriptions of the golem as a destructive creature draw more upon the story of Frankenstein than upon Jewish versions of the golem legend.

Scholars today continue to debate why the modern version of the golem legend became so closely identified with Rabbi Judah Loew of Prague. Although we can never know for sure, it seems that no other European rabbi could have served as a better candidate to become the paradigmatic creator of a golem than Judah Loew. Not only was his stature as a scholar, community leader, and national Bohemian hero unique, but his fame as a wonder-worker had been well established long before the creation of a golem was added to his already extensive curriculum vitae. Furthermore, what better setting than Prague to choose as the location for the creation of the golem? A city saturated with mystery and marvel, a city that in Loew's time served as the European capital of the world of the occult, Prague was the natural location for a tale of the supernatural.

In the sixteenth century, when Loew served as chief rabbi of Prague, the city was a major center both of alchemy and of science. The quixotic Hapsburg emperor, Rudolph II, governed from his castle in Prague. It is well known that the emperor was obsessed with science, magic, alchemy, and with Kabbalah, the Jewish mystical tradition. Historical documents tell us that Rudolph had a particular interest in the *Sefer Yetzirah*, The Book of Creation, that was believed to contain the formulas used to create life, including golems.

Documented eyewitness accounts testify to at least one meeting between Judah Loew and the emperor, on February 20, 1592. Precisely at a time when the emperor was most inaccessible, even to his ministers and courtiers, he summoned Rabbi Loew from the Jewish ghetto to the castle to meet privately with him. Such a rare and unusual event—a private audience of a rabbi with the emperor—could not go unnoticed. No one knows exactly what they discussed, but Loew's son-in-law who accompanied him to the palace records that the subject of conversation was "secret and obscure mysteries." Years later Loew's great-great-grandson recorded a family tradition according to which Loew had written and delivered a number of Hebrew amulets to the emperor.

Long before the legend of the golem was linked with Loew, Jewish and Czech legends were already circulating about Loew's wonder-working abilities. Many of these legends revolve around the historically documented relationship between Judah Loew and Rudolph II. For example, according to one legend, preserved both in Jewish and Czech literature, the very first meeting between Loew and the emperor did not take place when the emperor summoned Loew to the castle in February 1592 but years earlier on the Charles Bridge that straddles the Vltava River in Prague.

In the sixteenth century many edicts of expulsion were issued against the Jews of Prague. Often they were rescinded before taking effect. According to one legend, after Rudolph II had issued such an edict, Judah Loew sought an audience with the emperor to plead with him to cancel the decree, but the emperor's ministers refused to grant Loew an imperial audience. In desperation, Loew took matters into his own hands. He stationed himself on the Charles Bridge, vowing not to move until the emperor reversed the decree. Soon the emperor's splendid carriage approached the bridge, and his servants and soldiers ordered the rabbi to stand aside. But Loew stood his ground. The soldiers picked up stones and hurled them at Loew, but by the time they struck him the stones had magically turned to roses. As the emperor's carriage was about to trample Loew underfoot, the horses abruptly stopped of their own accord. The astonished emperor stepped out of the carriage and invited the rose-covered rabbi to the palace, where the decree was rescinded and a long-standing friendship was established between the emperor and the rabbi.

In another legend that struck the Bohemian imagination, a rose also plays an important role. According to this tale, Judah Loew had power over even the Angel of Death. Although the rabbi was in his nineties, neither his physical nor his intellectual powers had dimmed. One day his favorite granddaughter, Hava, brought him a rose to honor the oncoming Sabbath. Loew knew that the Angel of Death had hidden himself in the rose, and that if Loew smelled the rose the angel would claim his life. For this reason Loew refused to accept the rose. His granddaughter began to cry, and nothing Loew could say or do would stem her tears. Finally Loew gave in and accepted the rose. Later, as the Sabbath peace descended, Loew died. This legend inspired the Czech sculptor Ladislav Saloun in 1910 to create a sculpture of Rabbi Loew looking down at a young girl who is offering him a rose. The statue was hidden by the Czech underground during World War II and stands today next to the Old City Hall in Prague.

In Bohemia, Loew is revered as a national hero, perhaps the only rabbi in any nation to be so honored. Not far away from Saloun's statue, at the entrance to the old Jewish quarter in Prague, there stands another statue, a monument to Loew's golem. Today tourists continue to flock to the old Jewish quarter of Prague to walk in the steps of Judah Loew's golem. It is not uncommon to see tourists visit Loew's grave in the Old Jewish Cemetery for good luck before proceeding to gamble at Prague's many casinos.

Who better to become the prototypical creator of a golem than Rabbi Loew? What better time and place to set the legend of the golem than in sixteenth-century Prague? Loew's well-known mystical and magical prowess, amply described in both Bohemian and Jewish folklore, made him the perfect candidate to be assigned the task of catapulting a new version of ancient and medieval Jewish mystical traditions about the golem into the modern world. Once the golem legend and Loew were joined, beginning in the mid-nineteenth century, the golems that were supposed to have been created by all other rabbis and mystical masters were overshadowed and obscured by Rabbi Judah Loew and the golem of Prague. Once locked together forever by the web of folklore, Judah Loew and his golem became inseparable. Loew became the quintessential creator of the golem, and the golem he is said to have created became the model for all later golems.

Nascent forms of a new Jewish version of the golem legend featuring Judah Loew of Prague first appeared in Jewish, Czech, and German writings in the late 1830s and 1840s. In 1909 a full-blown version of the legend was published by Rabbi Yudel Rosenberg. This Hebrew work, *Nifla'ot Maharal, The Wonders of Rabbi Loew,* was claimed by Rosenberg to have been based upon a manuscript found in the Royal Library of Metz, part of which contained an eyewitness account by Loew's son-in-law who had assisted him in the creation of the golem. According to Rosenberg, the original manuscript from which he copied the story of Loew and the golem had been consumed by a fire that had destroyed the Royal Library. But Rosenberg's work turned out to be a forgery. There had never been a Royal Library in Metz. Nor had any library in Metz ever been destroyed by fire. Such a manuscript had never existed. Still, many Orthodox Jews today continue to believe in the historicity and authenticity of *The Wonders of Rabbi Loew.*

There is little doubt that Rosenberg was both an erudite scholar and a literary genius who was well acquainted with earlier versions of the golem legend. (Mordecai Richler, the eminent Canadian-Jewish novelist, was his grandson.) Rosenberg may even have had access to authentic medieval manuscripts about the golem that are now lost or unknown. He incorporated these classical teachings on the golem into his amplified account of the legend. Hence Rosenberg's version of the story of the golem and the rabbi of the Altneuschul, the Old-New Synagogue, was an old-new version of the legend of the golem, replete with old elements of the legend but also containing additions and embellishments drawn from Rosenberg's fertile imagination and from modern literature. Rosenberg is also the first writer to bestow a name on the golem, Yossele—a diminutive form of Joseph. Unlike the being created by Dr. Frankenstein, who is never given a name, Loew's golem is personalized and humanized.

Drawing upon the story of the golem of Elijah of Helm, Rosenberg describes Loew's golem as a servant whose powers threaten to run amok. Rosenberg casts Yossele Golem as a protector and savior of a physically threatened European Jewry. Of particular concern to Rosenberg is the reemergence in his own time of medieval traditions about the "blood libel"—the canard that Jews murdered Christian children to use their blood in certain Jewish rituals. Yossele Golem is described as uncovering plots against the Jews and punishing the perpetrators of the blood libel charge.

In one episode Loew emerges as a great detective. But scholars have demonstrated that Rosenberg's description of Loew as a detective was drawn from a story about Sherlock Holmes by Rosenberg's contemporary, Arthur Conan Doyle. Rosenberg also seems to have incorporated elements of other European literary works, such as the "Sorcerer's Apprentice," into his retelling of the golem legend.

In 1920, Hayyim Block translated Rosenberg's work into German. Bloch added some embellishments of his own to the story. In 1925 an English translation of Bloch's work appeared. Once available in German and English, and later in other Western languages, this new Jewish version of the legend of the golem—depicting it as a potentially dangerous being—was poised for dissemination beyond the confines of an exclusively Jewish readership. The story of the golem in its modern form was now positioned to infiltrate and influence European (and later American) arts, letters, and science. For this reason the Israeli scholar Joseph Dan has described Rosenberg's book as "the best-known contribution of twentieth-century Hebrew literature to world literature." Although it is a literary forgery, Dan considers *The Wonders of Rabbi Loew* to be a masterpiece of modern Hebrew literature.

From the first decades of the twentieth century, an obscure Jewish mystical tradition about the creation of life from inert matter has increasingly influenced literature, art, and film. A sixteenth-century rabbi, previously known only within limited segments of the Jewish community, has become a protagonist in a wide variety of genre of European cultural expression. From the vast number of legends and myths told by Jews throughout their long history in many lands, the tale of the golem has emerged as the most popular, pervasive, and influential of all postbiblical Jewish tales. Why?

Because like all enduring myths, the legend of the golem of Prague—though fiction and though set in a particular place and time—is a story that speaks to the human condition at all times and in all places, including our own—especially our own. The golem legend deals with enduring universal problems: the mystery of life, the nature of creativity, the appropriate uses of human power and creativity, the promise and the dangers of human tinkering with nature, the relationships we have with the artifacts we create to help and defend us but that also threaten to harm and destroy us.

The primary purpose of an enduring myth is not to entertain but to address the perennial questions that define our place in the world: who we are, where we come from, who we can become, where we find the wisdom to chart our future. The durability of the golem legend is sustained both by the nature of the questions that it poses and the responses it offers to those questions. In the chapters that follow, these questions will be examined. But before reflecting upon the implications of the golem of Prague for our "age of the golem," we must address the questions of why and how Judah Loew created his golem.

Chapter Three

Creating a Golem

A mélange of events set the stage for Judah Loew's creation of the golem of Prague. Among them were decrees of expulsion of the Jews from various lands by the once-protective monarchs of Europe, including Bohemia. The Jews of Spain, who had been prosperous and secure in Iberia for centuries, already had suffered the trauma of expulsion. The earthquake of exile that had devastated Spanish Jewry in 1492 sent aftershocks of insecurity to all corners of the Jewish world. Jews could not feel safe anywhere. For European Jews, their world had become a dangerous neighborhood.

As the new sixteenth century unfolded, new threats emerged. In Germany, Martin Luther called for the expulsion of the Jews from Germany and the confiscation of their property. In Vienna, sacred Jewish books were used to ignite public bonfires—would Jewish bodies be next? Suspected of being in a secret alliance with the Ottoman enemy of the declining Holy Roman

Empire, Jews were vilified as a fifth column in the European lands where they resided. The long dormant "blood libel" surfaced once again. The times required radical action. The Jews of Europe turned to Judah Loew, Kabbalist and alchemist, mystical master and wonder-worker, for protection.

Judah Loew knew that the threat was real, that the time for diplomacy and negotiation had passed, that radical action was now required. He recognized that the enemy would understand only force. Terror must be met with physical power.

Early in 1580, with Passover near, Loew knew that the threat of violence would escalate. The blood libel charges would return. Physical attacks would become more common. Marked as infidels by their enemies, his people would again be vulnerable. And so Judah Loew decided to summon all his magical and mystical knowledge to create a being—immense in physical prowess, of awesome size and strength—to defend his people from the inevitable assaults. This being—in many ways more than a man, in some ways less than a man—was the golem.

For years Loew had taught his disciples the intricacies of Jewish law and the secrets of the Jewish mystical tradition. He had instructed them in the perilous techniques of "practical Kabbalah," of magic and wonder-working. Now it was time for them to employ what they had learned. On March 20, 1580, before sunrise, Judah Loew went with his disciples, Rabbi Isaac Katz and Rabbi Sasson, down to the banks of the Vltava River. They stood together in the shadow of the Charles Bridge—where Loew had supposedly once dazzled the emperor with his magical abilities—preparing to do what they were not sure they could do: bring a golem to life.

In the preceding weeks they had rehearsed the obscure mystical texts that contained the recipe for creating life. They had fasted, repented, and prayed. They had purified themselves in the *mikvah*, the ritual bath, to purge their bodies and souls of sins and impurities.

On the eve of the twentieth day of the Hebrew month of Adar, they met at the Altneuschul in final preparation for giving life to a being that, until now, they had encountered only in obscure scholarly books. After making their last ritual ablutions, they studied for a final time the texts in which the mysteries of creation had been locked up for so many centuries. Then they donned white linen garments—as the high priest had done in Solomon's Temple—and walked together toward the river.

Judah Loew took his staff and drew the figure of a man on the river bank. He drew with the skill of an accomplished artist. Then he began to walk in a circle around the figure he had drawn while intoning the secret names of God found in the mystical treatises he had studied and taught for so many years. His two disciples followed him, circling the figure seven times counterclockwise, then seven times clockwise, while reciting the secret names of God and other magical formulas that promised to bring life to inert matter.

Judah Loew and his disciples spoke the letters of the most sacred word known, the mysterious Name of God, preserved for generations by the Kabbalist masters and passed in secret from master to disciple in centuries past. They intoned each consonant together, precisely pronouncing it with the appropriate vowel, exactly 310 times. For in Hebrew the letters are also numbers, and the letters corresponding to the number 310 spell the Hebrew word *yesh*, "that which is."

As they pronounced each of the letters, they envisioned the shapes of each of the consonants and vowels in the deepest recesses of their minds. As they circled the figure, each of the sacred letters they pronounced seemed to animate one of the limbs of the figure embedded in the river bank.

Loew took a cup of water and poured it on the chest and nose of the figure, and then he blew upon the spots where he had poured the water. He then recited the verse from the biblical account of the creation of Adam: "A flow would well up from the ground and water the whole surface of the earth—the Lord God formed man from the dust of the earth. God blew into his nostrils the breath of life, and man became a living being."

As Loew spoke, the figure became progressively detached from the earth. Hair grew on its head. Nails sprouted on its fingers and toes. Loew cautiously approached the figure, and under its tongue he placed a small parchment slip on which the words *Adonai Emet*, "The Lord is truth," were written. Loew knew that only the power of God, not his own power, could bestow life.

Breath now surged through the golem's nose and chest. Its eyes opened and sparkled in the sunlight of the day being born. A jolt of energy surged through its massive bulk. Firmly, Judah Loew commanded, "Golem, stand." And the golem rose from the ground, looking like anyone arising from a deep sleep. The golem nodded, showing Loew that he understood what had been said.

The body of the golem looked like Michelangelo's sculpture of King David—tall and strong, with muscles and sinews like fashioned marble, ready and able to confront any physical challenge that might lie ahead. Like golems before him, the golem of Prague was human in every way, except that it lacked the power of speech and the ability to procreate. It also lacked a human soul, having instead the kind of soul that most animals might have.

Named Yossele Golem by Loew, the golem, at Loew's commands, performed various tasks for which it had been created. Yossele was a servant in Loew's home, drawing water and chopping wood. He acted as a spy, infiltrating the Gentile quarter of Prague, seeming to be a country bumpkin to whom people would tell their secrets. He would watch and listen, then return to Judah Loew. Loew even prepared an amulet for the golem that rendered him invisible when he wore it.

By means of hand signals, the mute golem would report what he had learned. In particular he learned of schemes aimed at falsely implicating the Jews in the murders of Christian children, which were to be used to agitate the Christian mob against the Jews with charges of the blood libel. Based on the information gathered by the golem, Rabbi Loew and the golem would thwart such conspiracies and bring the criminals to justice. Thus pogroms against the Jews were prevented.

When attacks against the Jews of the Prague ghetto could no longer be averted, Yossele Golem became their defender, their champion, defeating any enemy with his physical might. Impervious to sword or flame, no physical element could harm him. No human being, except Rabbi Loew, could destroy him. In the demands of the moment, he could grow to enormous proportions. His very appearance could intimidate even the fiercest foe. The golem was physical power personified: he was the projection of physical and military force by a people who then had none.

Once the golem had completed his mission, once he had accomplished the tasks for which he had been created, the time came for him to be laid to rest. According to the legend of Loew and his golem, in the spring of 1590 Loew decided that his golem's mission had been fulfilled and that it was time to bring its life to an end. In most versions of the story of the golem, ancient and modern, longevity is rarely a feature of a golem's existence.

There are various traditional instructions about how to deactivate a golem. According to one, just as a golem is brought to life by the recitation of certain words and letters, it may be deactivated by reciting them in reverse. According to another tradition, the golem is brought to life by writing certain Hebrew words either on its forehead or on a piece of parchment placed somewhere in its body, often under its tongue. If the golem is activated by the Hebrew word *emet*, truth, it may be deactivated by removing the first letter of that word, leaving the word *met*, which means dead. Once the word *dead* appears on the golem, it reverts back to the elements from which it had been created. In various versions of the story of the golem of Prague, the golem is laid to rest by using one of these methods.

Unlike golems in ancient and medieval Jewish literature, modern golems are created for practical purposes—to serve as physical laborers, for example, or to defend innocent people from harm and violence. But as the golem legend was weaned away from its ancient and medieval origins, the golem was increasingly portrayed as a dangerous creature. Creation of a golem now became a prelude not only to the evocation of hope but to the anticipation of danger, not only a harbinger of redemption but an agent of destruction. The golem was transformed into a physical force both to be feared as well as admired: a slave who could become a master, a eunuch who could develop uncontrollable lust, a mute who could acquire the powers of speech and wax eloquent, a quaint curiosity who could run amok and threaten life and property.

In one popular version of the story of the golem of Prague, Yossele goes beserk and begins to tear up the Jewish quarter of the city, just at the start of prayers welcoming the Sabbath.

Judah Loew is summoned. He subdues the golem and takes it to the attic of the Altneuschul. There Loew removes the Name of God from its body, and the golem collapses into a huge pile of dirt. Unlike later versions of the golem legend, the golem here remains under the ultimate power and control of Rabbi Loew. Unlike Frankenstein's monster and contemporary renditions of the golem legend influenced by the story of Frankenstein, the golem in modern Jewish legend may become dangerous and destructive, but it always remains under the control of its creator. Sages like Judah Loew of Prague and Elijah of Helm, who are deemed wise enough to create a golem, are also considered wise enough to know when *not* to create one, and when to destroy one that already has been created.

Before he died, Judah Loew decreed that no one be allowed to ascend to the attic of the synagogue where the golem's remains reposed, except his successors as chief rabbi of Prague. In the eighteenth century one of his successors fasted and prayed for weeks in preparing to enter the attic. But when he began to climb the steps to the attic, he fainted. When he recovered consciousness, he forbade anyone ever to enter the attic to view the remains of the golem. Many years ago a friend of mine, the late Rabbi Ira Sud, who was a native of Prague, told me that in the 1920s when he was a boy, he and a friend climbed to the window of the attic of the Altneuschul synagogue to see if the remains of the golem were there. They looked inside and saw a large mound of red dirt covered with old tattered prayer shawls.

In Prague today there are those who believe that the golem still walks the streets of Prague at night. In any case, it is undeniable that even today the presence of the golem continues to pervade the city. Meanwhile, in the Old Jewish Cemetery in Prague, Rabbi Judah Loew rests in peace. But not his golem. The golem has been revived to live again in different ages, in different forms. Each time the golem legend is retold, the golem is brought to life once more. In the chapter that follows we will see how classical and modern versions of the golem legend have infiltrated and influenced European and American culture, of how the golem has become a metaphor for life in our times.

Chapter Four

Golems Among Us

Centuries ago, *golem* was a word known only to a small cadre of scholars who pored over recondite Hebrew texts. Today a search for *golem* on google.com yields hundreds of thousands of references. Throughout the twentieth century, and now into the twenty-first, the golem has become something of a celebrity, featured in novels, appearing in films and on television. Indeed, for many writers and artists the golem has become a metaphor for our times. For many scientists and engineers the golem serves as a symbol of current and future developments in robotics, genetic engineering, computer science, and artificial intelligence (AI).

Not only is the golem increasingly present in contemporary art and science, but art and science themselves have each been described as a golem. In their books *The Golem: What You Should Know About Science* and *The Golem at Large: What You Should Know About Technology*, sociologists of science Harry

Collins and Trevor Pinch portray contemporary science and technology as golems. They claim that this perspective offers a viable alternative to two equally mistaken views about the nature and implementation of science and technology. One view considers science and technology to be all good, the other as all bad. For some, science and technology are panaceas that promise to resolve the problems that beset humankind, including starvation, poverty, poor education, and illness. For others, science and technology represent ever-present threats to religious faith, human dignity, freedom, and beauty. For some, science is a knight-errant sent to combat ignorance and superstition. For others, science and technology are villains responsible for the proliferation of harmful radiation, weapons of mass destruction, the degradation of the biosphere, and profit-hungry technocrats.

For Collins and Pinch, however, science and technology are neither beneficent genies nor malevolent demons. They are golems. As these two sociologists put it, "The personality of science is neither that of a chivalrous knight nor pitiless juggernaut. What then is science [and its application, technology]? . . . a golem. A golem . . . is powerful. It grows a little more powerful every day. It will follow orders, do your work, and protect you from the ever threatening enemy. But it is clumsy and dangerous. Without control a golem may destroy its makers with its flailing vigour; it is a lumbering fool who knows neither his own strength nor the extent of his clumsiness and ignorance. A golem, in the way we intend it, is not an evil creature but it is a little daft. Golem science is not to be blamed for its mistakes; they are our mistakes. A golem cannot be blamed if it is doing its best. But we must not expect too much. A golem, powerful though it is, is the creature of our art and craft."

Like science, art too may be described as a type of golem. As early as the thirteenth century, Jewish mystics described representational art, in which images (especially of human beings) are depicted, as a kind of golem. More recently the Pulitzer Prize–winning novelist Michael Chabon has compared both the process and consequences of the novelist's art to the making of a golem. Like the creator of a golem, the novelist brings images to life one letter at a time, and is imperiled by his or her creation "as not only an inevitable, [but] necessary part of writing fiction." In his essay "The Recipe for Life," Chabon observes that "Much of the enduring power of the golem story stems from its ready, if romantic analogy to the artist's relation to his or her work. . . . As the kabbalist is to God so is a golem to all creation: a model, a miniature replica, a mirror—like the novel—of the world." Chabon's observations came as a reflection upon the process of writing his own novel *The Amazing Adventures of Kavalier and Clay*, in which the golem of Prague plays a "small but crucial role."

The 1978 Nobel Laureate in Literature, Isaac Bashevis Singer—himself the author of a children's book about the golem—also compared artistic creativity to golem-making. In Singer's words, "Writers felt in the legend of the golem a profound kinship to artistic creativity. Each work of art has the elements of a miracle. The golem-maker was, essentially, an artist. . . . Whenever I tried to write or create a story, I always felt as if I was witnessing a miracle. It was a miracle that I took part in. In my own ways, I too was creating little golems. . . . Artists create surprising golems which puzzle themselves and the people of their time."

In her scholarly study of the depiction of golems in contemporary art, film, and literature, Emily Bilski suggests that "In making a golem, the mystic takes inanimate material and infuses it with life. The magical act of creation is a powerful metaphor for the miracle of creativity and the creation of the art object. However, when a golem is brought to life to serve its creator, his power can grow to the point where he can no longer be controlled. The golem becomes a threat to those he was intended to help. Man's inability to control his creations is a theme developed in [modern] artistic interpretations of the golem legend. Once a work of art is produced, the artist loses control over his or her creation. . . . Power, whether scientific, political or artistic, once unleashed, cannot always be held in check."

Like Singer, the Nobel Laureate Elie Wiesel has written a children's book on the golem. Wiesel describes the golem as "the most famous creature in Jewish lore and fantasy." Of all Jewish stories since the Bible, and of all Jewish mystical ideas, none has become as well known or has been depicted so extensively and in so many genre of cultural expression—art, film, music, drama, poetry, novels, and short stories. Golems also appear in pop culture—rock music, children's games, comic books, and children's books, such as those by Wiesel and Singer.

In 1996, David Wisniewski's *Golem* was awarded the prestigious Caldecott Medal in children's literature. Reflecting on the golem legend, Wisniewski has written, "The story of the golem serves as a cautionary tale about the limits of human power. It has inspired the works of composers and authors. . . . The tale may even prove prophetic—as the fields of computer science, robotics, and gene manipulation advance, technological golems may arise in our culture. . . . In this allegorical fashion, Golem still lives."

In the early twentieth century, authors retold the golem legend in various literary forms: short stories, like those by the

Hebrew writers David Frischmann and Y. L. Peretz; plays, like the powerful, complex, but perplexing Yiddish drama *Der Goylem* by Halper Leivick; and, of course, poetry and novels. The first significant novel about the golem appears to be Gustav Meyrink's *The Golem,* published in 1915. A masterful surrealistic work set in a futuristic Prague ghetto, Meyrink's story seems to have been influenced more by Hindu ideas and the theosophical teachings of the popular Madame Blatavsky than by Jewish lore and Kabbalistic tradition.

In the 1970s and 1980s a number of golem novels appeared, for example Cynthia Ozick's *The Puttermesser Papers* (1987), a vastly expanded version of her earlier short story "Puttermesser and Xanthippe." Here a sexually and professionally frustrated New York City civil servant creates a female golem who saves New York from the chaotic state into which it has fallen. Eventually this golem is undone by her own lustful longings, which pose a threat to those around her. As a result, she is destroyed by her creator.

In the first few years of the twenty-first century a rash of novels on the golem have appeared—*The Procedure* by Harry Mulisch, Thane Rosenbaum's *The Golems of Gotham*, Nami Eve's *The Family Orchard,* and Frances Sherwood's historical novel, *The Book of Splendor.* As one reviewer remarked, there are now enough books to fill a "golem section" at Barnes and Noble. Contemporary literature has contracted "golem-mania."

Of the many poems composed on the golem, one may note the charming poem by the Argentinean poet Jorge Luis Borges. In response to Borges's poem, the gifted American poet John Hollander—himself a direct descendant of Judah Loew—composed "A Propos of the Golem."

In music and dance we have "The Golem Suite" by Joseph Achron, an opera by Eugene d'Albert that premiered in Frankfurt in 1926, and more recently a 1989 opera by John Casken. In a lighter vein there is "The Golem Shuffle" by Dan Gottshall.

The cinematographic arts discovered the golem as early as the age of silent films. Especially noteworthy are the films of the German director and actor Paul Wegener. His 1914 film was significantly surpassed by his now classic 1920 movie, *The Golem: How He Came into the World*, which Wegener directed and in which he also played the golem. Here the golem is portrayed as a lustful and destructive creature, more akin to Frankenstein's monster than to traditional Jewish versions of the golem legend. The 1935 Czech-French film *Le Golem* portrays the golem as a redeemer from the rule of autocratic tyrants. As the golem arises to liberate Prague from the rule of the imperial court, the film's motto, "Revolt is the right of slaves," is repeated by various characters. It is not coincidental that in the 1930s, with the spread of fascism, a renewed interest in the golem as liberator and redeemer appears in film, literature, and art. Throughout Europe, but especially among the avant-garde artists of Bohemia, fascism was correctly perceived as a profound threat not only to life but to free artistic expression. The golem was often evoked as a symbol of the liberation of the oppressed from a tyrannical oppressor.

More recently, American filmmakers have portrayed the golem as a liberator from less ominous forms of oppression. In 2001, Pete Hamill's best-selling novel *Snow in August* appeared as a movie on cable television. Here an eleven-year-old Irish-American boy and a rabbinic refugee from post-Holocaust Prague form an unusual bond. Based on traditions he hears from the rabbi, the boy creates a golem to protect himself, his widowed mother, and the rabbi from members of a local Irish gang. Unlike other versions of the golem legend, in this story the golem is not destroyed after he accomplishes his mission; nor does the golem pose a danger to society. Even the now cultic television series "The X-Files" featured an episode entitled *"Kaddish,"* about a rabbi who creates a golem for protection and revenge against local hoodlums. A 1995 short film starring Ed Asner, called *The Golem in L.A.*, describes how a golem is created to help save a synagogue from demolition by greedy real estate developers.

In his 1975 *A Psychohistory of Zionism*, the Israeli scholar Jay Gonen expands upon earlier understandings of the golem as a protector and savior of the Jewish people to offer an explanation of Zionism and the State of Israel. In Gonen's view, Israel is a political permutation of the golem legend. Like the golem, the State of Israel has been created to protect the physical safety of the Jews by means of military and physical might. For Gonen, whether Israel will become ultimately redemptive or destructive, a savior and protector or a danger to Jewish continuity, remains an open question.

In fact, not only the State of Israel but all nation-states can be described as golems. Like Rava's golem, a state may be viewed as an "artificial man." Like the golem of the Prague ghetto, it has been created to protect its citizens. In his famous seventeenth-century work of political philosophy, *The Leviathan*, published forty-two years after Judah Loew's death, Thomas Hobbes wrote, "Art goes further, imitating that rational and most excellent work of nature, man. For by art is created that great Leviathan called a Commonwealth or State—in Latin *civitas*—which is but an artificial man, though of greater stature and strength than the natural, for whose protection and defense it was intended. . . . The pacts and covenants by which parts of this covenant were first made, set together, and united, resemble that fiat, or the 'let us make man,' pronounced by God in the creation."

Much of the best artistic work that visually depicts golems has been produced as posters to advertise plays and films related to the golem legend, such as the French film *Le Golem*, Leivick's Yiddish play *Der Goylem*, and for children's books on the golem like those by Wiesel, Singer, and Wisniewski. And the costumes and props for these films and plays offer many intriguing ways of visually depicting the golem and his creator, Rabbi Loew. How these dramas have been enacted also provides valuable insights into the complex relationship between Rabbi Loew and his golem, between creator and creature.

Between the world wars and after World War II, numerous works of art presented images of golems as robotic and mechanical artificial human beings, addressing a growing apprehension about the implications of an increasingly mechanized society. Artists and writers used the golem to depict the fear of anonymity and the dissolution of individuality and creativity in modern life. The golem was chosen as an apt symbol of the individual as a machinelike automaton manipulated by nation-states and bureaucratic institutions. It became a metaphor for the dangers posed by the very developments in science and technology that had promised to redeem human beings from their physical and economic misfortunes. One such danger was that humans might become like the machines they had created: lifeless, emotionless, soulless automatons. The German writer and statesman Walter Rathenau composed a fable about a rabbi who replaces his aging, ugly, but virtuous wife with a beautiful female golem. Although the golem bears him a son, she remains devoid of emotion, even when their son dies. The rabbi then returns to his wife and begs her forgiveness.

In the 1950s American artists such as David Aronson used the golem motif to explore facets of the divine-human relationship, especially the features of golem-making as an act of communion with God. Aronson thereby returned to an understanding of golem-making deeply rooted in medieval Jewish mystical views of the golem. In the 1960s artists like Saul Steinberg explored the golem as a metaphor for the mystery of the artwork. Steinberg's drawing of the golem was accompanied by music released by Golem Records. In the 1980s, Toby Kahn reflected upon the artistic endeavor as a form of golem-making: "Every artist dreams that his or her work take on its own life, that it tell an independent truth. And yet in each creation there is an element of the golem, where beauty goes wild. Without a soul the piece will return to dust; too great a soul may lead to idol/icon worship." Like I. B. Singer, the artist Christian Boltanski sees art as magical, the product of an alchemist, a golem-maker. This insight is expressed in Boltanski's 1988 shadow installation pieces, of which *Golem* is an example.

Although writers and artists like Kahn, Boltanski, Singer, and Chabon explore the golem as a metaphor for artistic creativity, the artist Abraham Pincas goes one step further in his 1980s series *Golem and Tselem*. Pincas wants to involve the viewer as a participant in golem-making. He amplifies the display of his paintings with earth, simulated fire, controlled lighting and sound, and a recording of the chanting of Hebrew letter combinations used to activate the golem according to the works of the thirteenth-century Jewish mystic Abraham Abulafia. For Pincas, not only the artist but also the viewer can participate in the art of golem-making, particularly in the process of bringing art— inert matter—to life.

Beginning with Meyrink's novel on the golem, writers and artists have also used the golem as a symbol for the "dark side"— not only of the creative process and of the creator of golems but also of the human personality. Here the later modern identification of the golem as an irrational and dangerous creature is projected onto the nature of the human psyche. In many representations of the golem in today's popular culture, it is portrayed as evil personified. In *Superman* comic books, for example, one of Superman's archenemies is identified as the "Galactic Golem." A completely opposing representation of the golem is found in *Mendy and the Golem*, a comic book series sponsored by Hasidic Jews, wherein the golem acts like a Jewish version of Casper, the Friendly Ghost—a benevolent and helpful companion. Golems also play a role in such popular children's games as Pokemon and Dungeons and Dragons.

For decades, golems have been featured in science fiction stories, such as Avram Davidson's whimsical short story "The Golem." "Star Trek" enthusiasts have published the specifications of a twenty-fourth-century "Golem-class" starship whose sheer mass exceeds even the largest ships of the Federation of Planets' fleet.

If "today's science fiction is often tomorrow's scientific fact," as Stephen Hawking suggests, it should not be surprising that many of the speculative fantasies of science fiction have found their way onto the "to do" lists of scientists and engineers. Among them are those that refer to the golem. Some are described in the chapters that follow, but here are a few examples.

Norbert Wiener, the so-called father of cybernetics, launched a discussion of the implications of the golem legend for contemporary science and technology in his 1964 book *God and Golem, Inc.* Wiener described the machine as "the modern counterpart of the Golem." According to him, the golem legend anticipated many of the problems of the relationship between humans and machines, which Wiener identifies as one of the central concerns facing human society. A year later (1965), the reigning scholar of Jewish mysticism, Gershom Scholem, who had written a now classical study on the golem in Jewish mystical literature and tradition, delivered a speech at the installation of a new computer at the Weizmann Institute in Rehovot, Israel, in which he described the computer as a contemporary manifestation of the golem.

Today many computer programs and projects in robotics, as well as individual robots, have been named "Golem." One of many examples is "The Golem Project" at Brandeis University, dedicated to the "Automatic Design and Manufacture of Robotic Lifeforms." Besides its reference to the golem legend, this project employs the acronym GOLEM for Genetically Organized Lifelike Electro Mechanics. GOLEM is also used as an acronym for Geomorphically Orogenic Landscape Evolution Model— a university-based project to study topography.

There have been three major phases in the development of the golem legend. In the initial phase, forged in the writings of the ancient rabbis and the medieval Jewish mystics, the golem is created for no practical purpose. Rather, golems are brought into existence as part of a mystical and magical quest to achieve communion with God and to share in the creative rapture that God experienced when creating the world.

In the second phase, beginning with Elijah of Helm and crystallizing in various versions of the story of the golem of Prague, it is created for practical purposes—to be a personal servant, for example, or the protector of innocent people through the employment of its immense physical prowess. In this phase the golem's power is described as sometimes running amok and as posing a danger to its creator as well as to life and property. Yet the golem's power never extends beyond the point where its creator can no longer control it. When necessary, the golem's creator can bring the golem's rampage to a halt and its existence to an end.

In the third phase, as portrayed in a wide variety of twentieth-century European and American literature and the arts, the golem is presented as a malevolent creature, bringing inevitable tragedy and catastrophe as it evades the control of its creator. In this depiction the golem is more a facsimile of Frankenstein's creature than it is the scion of golems described by either classical or modern Jewish versions of the golem legend.

In the chapters that now follow, the teachings of the golem legend, as it developed in its first two phases, will be brought to bear on recent and anticipated developments in such controversial areas as bioengineering, reproductive biotechnology, robotics, artificial intelligence, and corporate ethics. It should become increasingly clear that the insights of the golem legend, embedded in classical and modern Jewish literature, offer a compelling and viable approach to the implications of new developments in science and technology.

With newspapers and magazines reporting stunning new advances in bioengineering almost daily, with ongoing public policy debates about cloning, with continuing attention to artifical intelligence and "artificial life," the story of the creation of life known as the golem legend is no longer a grandmother's tale. It has become a metaphor for our times as we confront the challenge of coexisting with the golems who now dwell among us in our biotech century.

Chapter Five

The Past Meets
the Present:
The Golem and
the Genome

       In the beginning was the word. Scripture describes God as creating the world with words: "Let there be . . . God spoke, and there was. . . ." Words create worlds, life, and golems. As we have seen, medieval commentators to the Talmud believed that Rava animated his golem with words. By employing a mystical method of letter combinations, Rava created a golem.

       For the medieval Jewish mystics, the twenty-two letters of the Hebrew alphabet, and especially the four letters of God's ineffable name—the Tetragrammaton, consisting of the Hebrew letters Yod, He, Waw, and He—if properly recited and skillfully combined, could be used to create a variety of life-forms. A flaw in the order of the letters, or an incorrect number of letters employed, would produce a flawed creature. Various letter combinations could be used to destroy as well as create.

Medieval Jewish and Christian philosophers, theologians, and scientists, such as Augustine, Gersonides, and Galileo, described God as the author of two books, Scripture and Nature. In this view, the theologian and the scientist share a common vocation: to decode the book written by God. For Jewish scholars this means decoding the Torah, written with the Hebrew alphabet, whose letters are pregnant with creative potency. Indeed, according to the medieval Jewish mystics, the Torah is one long name of God. If the adept mystic could discern the proper primordial order of the letters of the Torah, the letters of God's long name, he could work wonders. He could crack the code of creation.

With the sequencing of the human genome, these abstruse medieval mystical notions have taken on new meaning and relevance. Like the Torah, the genome, sometimes called the Book of Life, consists of sequences of letters comprising a very long text. Decoding this book enables us to discover many of the secrets of life. If we could master the art of combining the letters that comprise this book, we could work wonders.

Mastering the combinations of the four letters of the Tetragrammaton can help the mystic penetrate the mysteries of life and offers the potential for creating new life-forms. Similarly, four letters are used to represent the code of the genomic Book of Life. Each of these four letters, GATC, represents one of the four nucleotide bases in DNA: guanine, adenine, thymine, and cytosine. The genome is a long book consisting of a variety of combinations of these four letters. Learning how to read, decode, and understand the various sequences of these letters may provide a key to understanding life and to creating new life-forms. In his inventive novel *The Procedure* (1998), the Dutch novelist Harry Mulisch compares Judah Loew's creation of a living golem from inert dirt by employing the Kabbalistic method of combining and recombining Hebrew letters to the attempts of contemporary scientists to bring inert matter to life by the combinations and manipulations of the nucleotide bases of DNA represented by the four letters GATC.

In Jewish law, if a Torah scroll has even one defective letter or word, it is considered unfit for ritual use. Sometimes, however, a skilled scribe can correct the mistake. Similarly, a single misspelling, a mistaken duplication of a letter, or the absence of a letter in the chain of letters representing the human genome with regard to a particular gene, can indicate the presence of or predict the onset of a devastating disease. Sometimes, however, either now or in the future, developments in gene therapy may be able to repair such a mistake.

In the legend of the golem, the *liber scripturae* and the *liber naturae,* the Book of Scripture and the Book of Nature, converge. The golem legend teaches that, like the golem, life has been engendered by the permutations of letters. As the verse in Psalms reads, "Your eyes saw my unformed mass (Hebrew: *golmi,* literally: "my golem"); they were recorded in Your book." Commenting on this verse, the Talmudic rabbis describe how God created Adam, a golem, and showed him the book of the future of humankind. Later God infused a soul in him, transforming Adam-Golem into Adam-Human. In a sense, Adam-Golem read the story of human evolution and history before it occurred. Conversely, the publication of the "book" of the human genome offers us a retrospective look at the history of human evolution. It shows how we evolved to become human beings.

Just as Scripture consists of combinations of the same letters of the alphabet, so life is composed from the same nucleotides represented by the letters GATC in a variety of sequences and combinations. The discovery of DNA and the sequencing of the genome has demonstrated that all life on earth is constructed from the same biochemical elements. Human beings are made of the same "stuff" as bananas, frogs, mice, and chimpanzees. In fact about 25 percent of our genes are about the same as those of a banana, about 50 percent as a worm, about 60 percent as a fruit fly, about 90 percent as a mouse, 98.5 percent as a chimpanzee, and 99.9 percent as another human being. In other words, we are 99.9 percent the same as everyone else, different from one another in only one in each thousand of our DNA bases. Yet a single base—a missing letter or a wrong letter—can mean the difference between life and death, health or illness.

The sequencing of the human genome was a major scientific breakthrough. But it may also be seen as the most recent in a chain of scientific developments, beginning in the sixteenth century, that have progressively diminished the stature and place of human beings in the cosmos and on our own planet. Before the Copernican Revolution our world was viewed as the center of the universe, and human beings were considered the pinnacle of God's creation. Since that revolution we have had to accept the sobering fact that our world is not the center of the universe or even of our own solar system. Our Earth is merely a small planet orbiting a rather mediocre star that is part of an enormous galaxy. Even our galaxy is but one of many, and is apparently less remarkable than many other galaxies.

During the Middle Ages, human beings considered themselves to be the most important species on earth, perched only below God and the angels on the "Great Chain of Being." But with the inception of Darwin's theory of evolution in the nineteenth century, and the acceptance of the scientific claim that the human species has descended from lower life-forms, human beings were dethroned from their position as the crown of God's creation. More recently, revelations of the genome have further deflated human claims to biological uniqueness and superiority. We now know that human life at its base is no different from other terrestrial life-forms. Our genes are made up of the same components as those of fruit and fruit flies. How can each human being claim to be "special" or unique when he or she is genetically similar to a mouse and 99.9 percent identical to every other human being who has ever lived?

This deflation of human stature has led to two seemingly dissonant views among scientists. On the one hand we find scientists rejecting traditional religious views of humans having been created in the image of God and therefore possessing a uniquely intrinsic sanctity of life. Beginning with Carl Sagan in the 1970s, scientists have considered it "speciesism" for human beings to posit their preferential superiority over other forms of life. Sagan compared the human prejudice for the superiority of humankind over other life-forms to various forms of racism. More recently, the genomic research that revealed a common genetic and chemical foundation for all life on earth has led some prominent scientists and philosophers to describe those who oppose trans-species gene transfers as advocating a now obsolete, anti-scientific, religiously based form of speciesism. Some eminent philosophers, such as Princeton's Peter Singer, consider those who oppose even interspecies sexual relations as being guilty of speciesism. And cutting-edge scientists in robotics and related fields have accused those who resist the claim that machines may be considered as having human status, that humans ought to merge with machines (become cyborgs) and eventually *become* machines, as advocating speciesism vis-à-vis mechanized life-forms.

On the other hand, while calling for a new human humility and for humankind to accept its scientifically established deflated stature in the natural scheme of things, scientists simultaneously assert that nature is flawed, and that they are both empowered and obliged to fix it through technology. The same scientists who remind us of how much we have in common with bananas and mice want to take it upon themselves to improve nature. They have implemented a program of re-creating and re-engineering life on our planet, including human life, through the manipulation and recombination of elements of DNA—the building blocks of all life on earth. This enterprise is known as "recombinant DNA technology" or "genetic engineering."

At the risk of oversimplification, the history of science can be divided into three phases. In its earliest stage, in ancient and medieval times, science was primarily concerned with the question, What is nature? Beginning about the seventeenth century, the question became, How does nature work? In recent decades a new question has become paramount: How can we alter and improve nature to serve human purposes? This is not to say that contemporary scientists are no longer concerned with the nature and functions of natural phenomena. For example, astrophysicists continue to investigate what the universe is and how it works. But in many areas of science, especially in the biosciences, a major focus of attention is how we can improve upon what nature provides.

For many people, particularly members of various religious communities, this new scientific attitude toward nature is considered immoral, unnatural, and a reprehensible usurpation of divine prerogatives. But for actual and legendary creators of golems, like Judah Loew of Prague, this approach is neither new nor problematic, from either an ethical or religious perspective. Even in the sixteenth century, Judah Loew wrote, "Everything that God created requires repair and completion." In this view, creation is a process initiated, but not completed, by God. Human beings have a divine mandate to act as "God's partners in the work of creation," to improve upon the work that God initiated with the creation of the world. God provided raw materials for us to cultivate and develop. A rabbinic parable puts it this way:

Once, a king had two servants. One morning he gave each servant a pile of flour, a pile of flax, and a pile of grapes. Then the king left the palace for the day. Upon returning in the evening, he was tired, hungry, and thirsty. He called for the first servant, who said, "Your majesty, all day I have stood guard over the flour, the grapes, and the flax that you gave me this morning. Nothing has been stolen, nothing has been consumed." And the first servant returned the flour, the grapes, and the flax to the king exactly as they had been received that morning. Then the king called for the second servant. "Where is the flour, the grapes, and the flax that I gave you this morning?" asked the king. Said the servant, "All day I have been laboring to satisfy your majesty's hunger and thirst upon your return to the palace. I have kneaded and baked the flour into bread for your majesty to eat. I have made the grapes into wine for your majesty to drink, and I have woven the flax into a linen tablecloth upon which to set the wine and the bread that I have prepared from what your majesty has provided." The king sat at the table, which was covered with a beautiful linen tablecloth, and refreshed himself with the wine and bread prepared by the second servant. When he had finished eating and drinking, the king rewarded the second servant and punished the first.

This parable represents how the Talmudic rabbis understood the biblical verse (Genesis 2:3) that describes the world as having been "created to be made." From this perspective, the human mission is not necessarily to accept nature as it is but to try to improve upon it, to develop its raw materials for beneficial human purposes, to repair and improve upon God's initial creation. According to this parable, such activities are not a usurpation of God's sovereignty, not an "unnatural" activity, not "playing God," but rather a means of serving the Creator. Here we see how scientific and technological developments, whether they are the creation of golems or of genetic engineering, need not conflict with religious beliefs or sensibilities. Such activities are not prohibited but encouraged as long as they serve legitimate purposes. This does not mean we have carte blanche to do anything we want to do or are able to do with the world. Activities that do not "repair" or "complete" nature are discouraged. Activities that can wreak harm and destruction are prohibited. The human mission is to repair and improve the world and humankind, not destroy them. Technological ability must be tempered by moral wisdom.

In the modern period, science has moved away from questions of meaning and morality that preoccupied ancient and medieval scientists, focusing primarily on matters of "know-how" and technology. Issues of human purpose, wisdom, and ethics have been relegated to philosophers and theologians. For some theologians, it is precisely the application of morality and wisdom to the use of advanced technology that bears witness to the resilience of claims to human uniqueness and dignity in the face of their deflation by modern science. Scientific theories and discoveries about the place of our world in the cosmos, and the stature of human beings in our biosphere, need not entail a deflation of the dignity or sanctity of human life. For such theologians, both the belief in the intrinsic sanctity of human life and the affirmation of the human mission to repair the world derive from the claim that human beings, who are created in God's image, serve as "God's partners in the work of creation." As Anna Foerst, who served as the official theologian at MIT's artificial intelligence laboratory, put it, engineering can be a form of prayer, and the new technologies may be considered as a "sign of people trying to participate in God's creativity."

In the early 1970s, when scientists developed a technique for moving DNA from one organism to another, thereby creating new forms of life, recombinant DNA, or genetic engineering, was born. By combining and recombining the fundamental biochemical elements of life, many new life-forms could be brought into existence.

Recombinant DNA technology is the artificial manipulation of genetic material to gain a desired result. Often used synonymously with terms like "genetic engineering," "biotechnology" or "bioengineering," recombinant DNA technology can be applied to plants, animals, and humans as well as across species. As the basic components of all life, elements of DNA can be artificially added to, deleted from, or rearranged within an organism, or transferred from one organism to another. These recombinations of genetic material can alter an organism's form, function, and nature. Genetic engineering can be used for benevolent reasons such as curing and treating disease, for malevolent reasons such as creating weapons of mass destruction, and for reasons that might be considered frivolous—like creating fish with implanted genes from glowworms that "light up" in a darkened fish tank.

The introduction of recombinant DNA technology has been greeted with an outpouring of enthusiasm as well as anxiety. The quandary is how to develop these new techniques for peaceful and beneficial means while preventing their employment for lethal and destructive ends, and how to alter nature without irreparably harming it. To date this dilemma has not yet been resolved. Like much of modern technology, genetic engineering is a double-edged sword: it offers horrors as well as blessings.

The initial reaction to the inception of genetic engineering was fear. In 1976, anxious about the creation of new bio-organisms at labs in Harvard and MIT, the city council of Cambridge, Massachusetts, where those labs were located, imposed a moratorium on recombinant DNA research. But when beneficial uses of the new technology were demonstrated with the creation of genetically engineered insulin in 1978, public anxiety declined, and public opinion began to shift from fear-driven opposition to cautious acceptance. As long as DNA-related technology was perceived to be beneficial to the human condition, it received public support.

In 1987 DNA "fingerprinting" was used for the first time in forensics and in paternity cases; by 1998 it was used to help impeach a president of the United States. In the 1980s genetic testing for predispositions to certain diseases was first developed. In 1985 the first transgenic plant was produced—it manufactured its own insect repellent. Also in the 1980s, transgenic plants and animals were created that produced pharmaceuticals useful in treating a wide variety of human diseases and conditions.

It was the 1980 decision of the U.S. Supreme Court in the case of *Diamond v. Chakrabarty* that catapulted into the public square the ethical, legal, and social policy issues related to the creation of artificial life-forms. Until this case the creation of artificial life had been a matter largely limited to the domain of science, legend, and theological discourse. After the decision in *Diamond v. Chakrabarty*, the biotech revolution rapidly accelerated, and with it public discussion of the ethical and social implications of developments in science and technology expanded exponentially. As science and technology created new artificial life-forms, observers became increasingly preoccupied with the propriety and implications of these creations.

In March 1980, exactly four hundred years after Rabbi Judah Loew of Prague was said to have created an artificial life-form called a golem, the Supreme Court heard arguments about the patentability of an artificial life-form. In dispute was whether Dr. Ananda Chakrabarty, a microbiologist now at the University of Illinois at Chicago, should be granted a patent on a bacterium that in 1972 he had created in his laboratory. Unlike Rabbi Loew, Chakrabarty had not used Kabbalistic magic to create this new life-form. Rather, he had employed a then newly emerging technological magic known as genetic engineering. As Arthur C. Clarke, author of *2001: A Space Odyssey* and other works of science fiction, once wrote, "Any sufficiently advanced technology is indistinguishable from magic."

Using recombinant DNA technology, Chakrabarty had combined segments of DNA from existing bacteria to create a new bacterium with properties found in no known naturally occurring bacterium. In effect he had created a life-form that nature had not created, a new species of living being.

The plaintiff in *Diamond v. Chakrabarty* was Sidney Diamond, then U.S. commissioner of patents. His goal was to reverse a decision by the Court of Customs and Patent Appeals that had granted a patent to Chakrabarty for B. cepacia, the new bacterium. This microorganism had the ability to break down crude oil into simpler substances that could serve as food for aquatic life, a property found in no known naturally occurring bacteria. It could help clean up toxic spills by degrading environmental pollutants such as oil and Agent Orange. Thus it had considerable commercial and ecological value. Opponents of Chakrabarty's creation warned that pranksters could drop his bacterium into people's gas tanks, or that agents of oil companies could drop it into oil refineries to consume the competition's oil reserves.

Chakrabarty had applied to the U.S. Patent Office for patents both on the process of creating the bacterium and on the bacterium itself. A patent was initially granted for the process, but not for the organism. Chakrabarty then appealed to the Court of Customs and Patent Appeals for a patent on the organism.

The Patent Office had denied Chakrabarty's petition because precedent held that "products of nature," such as bacteria, are "living things" and are therefore not patentable. Chakrabarty claimed, however, that his bacterium was not a product of nature. It had not existed in nature until he created it in his laboratory. On appeal, Chakrabarty's position was upheld. On behalf of the Patent Office, Diamond then appealed this unprecedented decision to the Supreme Court. On June 16, 1980, in a five-to-four decision, the Court affirmed the split ruling of the Court of Appeals in favor of Chakrabarty. Writing for the majority, Chief Justice Warren Burger held that "a live, human-made microorganism is patentable subject matter under statute providing for the issuance of patent to a person who invents or discovers 'any' new or useful 'manufacture' or 'composition of matter.'" According to Justice Burger, "the relevant distinction is not between living and inanimate things" but rather between naturally existing and humanly fabricated inventions. Since Chakrabarty's bacterium was not "nature's handiwork" but a product of "human ingenuity and research," it was deemed patentable. With this decision, it became possible for a single human being or an individual corporation to secure monopolistic property rights over a living species. A precedent had been set for the future patenting of bioengineered plants, animals, and even human tissue and genes.

The Supreme Court's decision sparked a debate that continues to polarize American public policy. Advocates of genetic engineering laud its present and potential benefits; detractors emphasize its inevitable dangers. But once the products of bioengineering were deemed patentable, enormous funds became available for investment, research, and product development. Biotechnology became a growth industry.

As biotechnology develops further, so will its presence in every aspect of our lives, from conception to death. Genetic engineering currently affects the foods we eat, the pharmaceuticals we ingest and inject, and the medical care we receive. As this technology unfolds, it will affect the way we reproduce, the kind of children we have, our understanding and treatment of disease, the way we wage war, the way we defend ourselves from our enemies . . . and much more. Where the twentieth century has been called the "age of physics," the twenty-first century is already being called the "biotech century."

The biotech century of today and of a rapidly accelerating tomorrow is one in which:

The foods we eat and the pharmaceuticals available to us to fight and prevent disease will be increasingly produced by techniques in bioengineering.

New artificial life-forms will be produced in greater numbers.

Interspecies genetic transfers will escalate—for example, human genes will continue to be infused in plants and animals, and animal genes might be implanted in human beings.

Trans-species organ transfers, such as the transplanting of pig organs in human beings, will increase. For instance, human beings with failing hearts will receive either parts of pig hearts or whole pig hearts to sustain them.

Gene therapies will become more prevalent, both for the treatment of disease and for the enhancement of certain characteristics. For example, certain genetic therapies will be provided before birth, to fetuses; others after birth; some with "artificial chromosomes."

Asexual human reproduction will become more common as reproductive cloning becomes prevalent.

Children will be increasingly conceived in vitro and eventually may be gestated in artificial wombs (as in Aldous Huxley's 1931 novel *Brave New World*).

. . . And more.

These kinds of developments, and others not yet conceived, may offer us a world in which:

The genetic structures of organisms and their descendants, including human beings, are permanently altered through genetic engineering.

Biology and engineering merge into bioengineering.

Distinctions between organic life and machines become more and more ambiguous.

Distinctions between natural and artificial life become both ambiguous and irrelevant.

People interact more with machines than with other human beings; indeed, people establish and maintain more "relationships" with machines than with other human beings.

. . . And more.

Imagine a world in which the normal human life span is 150 years, where worn-out vital organs can be replaced by spares, where after your death you retain consciousness for eternity in cyberspace, where nanotechnology enables you to transform a plastic bottle into a filet mignon for you to share with your android spouse. Imagine a society governed not by national governments but by global corporations, where law enforcement officers have bulletproof chests created by genetic engineering, where transgenics has provided athletes with the sight of an eagle or the swiftness of a cheetah, where nanorobots flow through your bloodstream diagnosing medical problems and making repairs, where reproductive cloning and artificial wombs have become the preferred manner of having children, and where sexual relations have become a virtual reality. Scientists anticipate that within a century these features will characterize our society and our world.

Now imagine a world in which goats produce silk in their milk that can be used to manufacture bulletproof vests, where genes from a fish implanted in a rabbit make the rabbit glow in the dark, where a container of genetically engineered bacteria could wipe out the population of a major city, where computers are embedded in your clothes and in your body, where most of the food you consume is genetically altered, where computers write poetry and music and defeat chess masters at their own game, where frozen embryos can be implanted for gestation and birth decades after their initial fertilization. Imagine a society in which your genes, tissues, and blood have been patented by large corporations, where your known genetic predispositions can affect your employability and choice of a spouse. Such features already characterize our world and our society.

Recent and anticipated developments in science and technology—especially biotechnology, computer science, robotics, and "artificial intelligence"—will compel us to confront and reevaluate our thinking about fundamental philosophical, theological, ethical, and public policy issues. These include the nature of life, the creation of artificial life, the nature of nature, the nature of human nature and identity, the relationship of human beings to nature, the relationship of human beings to the artifacts they create, the nature and limits of human power and creativity, and moral responsibility for the entities we create. How we respond to these daunting issues will affect the kinds of human beings we become, the society in which we live, and the public policies that govern our lives and the lives of those who come after us.

Even such staples of American life as "motherhood and apple pie" are not immune to the scientific and technological realities that are already upon us. Mom's apple pie may look and taste the same as it always has. But today, more likely than not, its key ingredients—apples, wheat, and sugar—have been "genetically modified." Just as mom's apple pie may not be what it used to be, neither may be "mom." Because of developments in reproductive biotechnology, we can no longer assume that conception has taken place as it always has. "Mom" may be an egg donor other than the "mom" who gestated us in her womb or gave birth to us or raised us, or all of the above.

With the inception of bioengineering and with advances in computer technology, even the most fundamental of all things—*life*—is suddenly not as simple as it used to be. Today one can say about life what the Supreme Court said about pornography: we cannot precisely define it, but we know it when we see it. Or at least we used to think so. Now, as new life-forms increasingly populate our world, we cannot always be sure that we know life when we see it. Even scientists are no longer sure of what life is.

Many of us remember how the late Carl Sagan explained the universe to us in his television series "Cosmos." Yet even this eminent scientist had to admit that he and his fellow scientists could not agree on a definition of life. Writing about "Life" in the *Encyclopedia Britannica,* Sagan confessed that "a great deal is known about life. . . . Yet despite the enormous fund of information that each of the biological specialties has provided, it is a remarkable fact that no general agreement exists on what is being studied. There is no generally accepted definition of life."

Although scientists cannot agree on what life is, they nonetheless talk about "artificial life." In recent years many "artificial" or humanly designed life-forms have been created. Some are organic, as we are; some are not. Computer-generated programs have created new life-forms that inhabit the realm of cyberspace and "live" in silicon chips. We currently coexist with an increasing number of artificial life-forms, with a world full of new varieties of golems.

At a landmark conference held in Los Alamos, New Mexico, one of the sites where the atomic bomb was developed, scientists gathered in September 1987 to establish a new science they called "artificial life." At that conference the physicist James Doyne Farmer discussed the implications of this new science. He predicted that "Within fifty to a hundred years a new class of organisms is likely to emerge. These organisms will be artificial in the sense that they will originally be designed by humans. However, they will evolve into something other than their original form; they will be 'alive' under any reasonable definition of the word. . . . The advent of artificial life will be the most significant historical event since the emergence of human beings."

Farmer was not referring to life as we know it—that is, organic life. He was referring rather to computer-generated simulations of life-forms. He was not talking about organic matter but about information produced by a computer—not about life formed *in vivo* or *in vitro* but about life formed *in silico*, in silicon. Much of what used to take place in laboratory test tubes now takes place on computer programs. The creation and evolution of many artificial life-forms now takes place on computers. In a very real sense, much of contemporary science has become a branch of computer science.

In the 1970s, more than a decade before the term *artificial life* was coined, new artificial forms of organic life were being created by available techniques in genetic engineering. Such life-forms—like Chakrabarty's new bacterium—were "artificial" in the sense that they were humanly made and had not previously existed as part of nature. But, as we have seen, centuries before the use of recombinant DNA to create new life-forms, and before the existence of computer-generated "artificial life," Jewish mystics and jurists already had been preoccupied with the creation of artificial life. Their discussions coalesced around the creation of the golem. For those conversant with classical and modern reflections upon the implications of golem-making, many of the earlier noted philosophical, theological, social, and ethical challenges are familiar and will continue to confront us in our "biotech century."

Just as the synagogue where the golem of Prague is supposed to repose is called the Old-New Synagogue, the issues raised by the golem and by biotechnology are both old and new. The challenge before us is to discern how to apply the accumulated wisdom of the past to the perplexities of the present; how the old can speak to the new; what creators of golems in ages long past have to say to the creators of the contemporary golems that now dwell among us. No activity characterizes the biotech century as much as bioengineering. It is to the topic of bioengineered organisms, to "organic golems," that our discussion now turns.

Chapter Six

Organic Golems: Frankenfood and Designer Genes

Immediately after recounting the story of how Rava created a golem, the Talmud tells the tale of how Rabbi Hanina and Rabbi Oshaya created a small calf by magical and mystical means. Being poor, these rabbis had no food with which to celebrate the Sabbath, so each Sabbath eve they would create a calf which they would then prepare for their Sabbath meal. Here we have an anticipation of the creation of genetically modified food (GMF)—the creation of food utilizing the best available technology.

In recent years much of the public debate about genetic engineering has centered on food. Since we are what we eat, and since the food we consume often relates to who we are— medically, culturally, ethnically, and religiously—it is not surprising that genetically modified food has become a pivotal issue in the global public policy discussion about the employment of recombinant DNA technology. For example, people can readily understand why diabetics would prefer safer and cheaper insulin produced by recombinant DNA technology. Even animal rights activists can appreciate that synthetic insulin grown in a yeast-based culture is preferable to extracting insulin from the pancreases of tens of thousands of cows. Yet people remain skeptical about claims that genetically modified food is safe to eat.

Today, whether they know it or not, most of the food consumed by Americans has been genetically modified in some way. Recent studies reveal that more than 60 percent of all food produced in the United States has been genetically modified in one way or another. Many of the plants, fish, and animals we consume have been altered by genetic engineering. Tests by consumer groups have found altered DNA in foods ranging from hamburgers to soy burgers, from milk to dog food, from tomatoes to cake mixes, from breakfast cereals to smoked salmon.

Weeds and insects are two of the greatest enemies of food crops. They destroy and inhibit the growth of plants produced for human and animal consumption. Many genetically altered crops have had genes inserted to provide them with the ability to contend with these natural enemies. These genes enable such plants to tolerate herbicides and resist insects. Rather than killing weeds as well as damaging crops, herbicides can now be used to kill weeds but leave the crops intact. Insect resistance provides crops with the ability to produce their own insecticide, killing or injuring predatory insects that otherwise would damage them. These genetic modifications aim at producing more resilient and more abundant crops. Another important result is the alleviation of the need to use massive amounts of dangerous chemical insecticides that harm plants, animals, and humans. Early ecologists, including Rachel Carson, had warned of the many dangers of using insecticides such as DDT to protect crops. While no longer legal in the United States as an insecticide, DDT is nonetheless still used in other countries where it continues to cause disease in animals and humans.

Crop protection is a major goal of the genetic alteration of plants. Besides making crops herbicide tolerant and insect resistant, genetic modifications have been used to help protect plants from the effects of other natural enemies such as plant diseases caused by bacteria, viruses, fungi, and insects. In Hawaii, for example, genetic engineering saved the papaya industry from devastation when the papaya crop was threatened by a viral infection. Another natural enemy of crops is the weather. While plants cannot be protected from all types of harmful weather, some can be made more resilient against certain types of weather conditions, such as frost. Certain plants, like citrus fruit, are highly vulnerable to onsets of cold weather. It may now be possible to transfer genes from frost-resistant animals, such as arctic fish, to plants in order to counteract the otherwise devastating effects of cold, frost, and snow. It may also be possible to engineer plants whose own genes can be manipulated to withstand the cold.

Besides plant protection, genetic modification of plants aims at better crop quality and nutrition. Many of the millions of people throughout the world who suffer from vitamin deficiencies live in poor countries where rice is the primary food. Now "golden rice," enriched with Vitamin A, has become available to feed these people. By inserting genes from daffodils and elsewhere into rice, the new plant contains otherwise unavailable

Vitamin A. It has been estimated that "golden rice" has the potential of preventing blindness and/or saving the lives of one million children who might otherwise go blind and/or die from Vitamin A deficiency. Efforts are now under way to modify plants genetically in order to produce additional edible vitamins as well as edible vaccines. Various foods may become hybrids of food and pharmaceuticals. Children may soon be able to take their vitamins and their "pills" simply by finishing their dinner.

Scientists are working to produce genetically modified foods aimed at lowering blood pressure, inhibiting the onset of diseases such as arthritis, reducing the need for harmful chemical additives, and producing antibodies to fight infections. In addition, plant genetic engineering will be increasingly employed to degrade pollutants, to detoxify soil and water, and to monitor soil contaminants (as was done following the 1986 Chernobyl nuclear power plant disaster). There are plans to produce biodegradable plastics from bioengineered plants, which would replace the manufacture of plastics from nonbiodegradable petroleum-based materials that increase pollution.

Plant biotechnology is also beginning to address consumer preferences. In coffee, genes that express caffeine can be suppressed, making it possible to grow decaffeinated coffee beans rather than sacrifice taste, quality, and aroma by having to decaffeinate the beans. Soy milk that tastes less like soy and more like milk will become available. Milk products such as ice cream for lactose-intolerant consumers will be found on grocery shelves in the near future.

In fish farming, genetic engineering has led to the production of bigger and better fish. They have been genetically reprogrammed to produce growth hormones the year round instead of only in warmer summer months, or growth hormones from larger fish species have been implanted into smaller species of fish. As a result, certain species of fish, such as salmon, can reach market size in less than half the usual time. Growing faster, needing less tending and resources, such fish have become more plentiful and cheaper by the pound to raise.

Cattle that produce leaner and healthier meat are being produced through genetic engineering and cloning techniques. Genetically modified chickens are bred to maturity and are available for consumption much earlier than before. Israeli scientists have even produced a featherless chicken that does not require its feathers to be plucked before being processed for food.

Farm and other animals are not only being genetically modified to produce food but to manufacture pharmaceuticals. Transgenic animals are producing proteins in their milk, eggs, blood, and urine which have been used to make drugs that can help diagnose and treat a wide variety of human diseases and medical conditions, including cystic fibrosis, intestinal infections, hemophilia, strokes, heart disease, and some forms of cancer.

Yet despite the many obvious benefits of genetic engineering, strong opposition to its past and potential development continues. Nowhere is this more evident than in the ongoing public policy debate over genetically modified foods. The controversy is not theoretical but deadly serious. When the United Nations sent food in the summer of 2002 to alleviate starvation in Zambia, the food was refused because it contained genetically engineered elements. People chose to die rather than consume genetically engineered food.

Americans eat bioengineered food, and its use is expanding exponentially in countries like China. Nonetheless most Europeans strongly oppose the production and consumption of modified food. Why the ruckus?

Until the late 1990s genetic engineering was largely viewed as something done by scientists in laboratories, with little daily impact on people's lives. Discussions over the propriety and uses of genetic engineering were often limited to ethicists and theologians. But as the twentieth century ended and the new millennium began, a confluence of events extended the debate about genetic engineering into the public arena.

In the 1990s, the first commercially available genetically engineered plant, the FlvrSavr tomato, became available in American markets. Genetically modified to inhibit it from softening, the FlvrSavr tomato, which was more expensive than other tomatoes, was a commercial failure. Although in 1994 the Food and Drug Administration proclaimed the FlvrSavr to be safe, the FDA did not require it to be labeled as genetically modified. In the following years, massive commercial production of genetically modified foods appeared on grocery store shelves without being labeled as such. When consumer advocacy groups revealed that much of the food Americans were consuming was genetically modified, the controversy over genetic engineering literally struck home. People began to wonder and to worry. They wondered why they were not told that their food had been genetically altered. They suspected that something was being hidden from them. They were concerned that genetic tampering with their food "behind their backs" might pose a danger to themselves and their families. They sensed a cover-up.

In 1999 journalists in Great Britain revealed that the British government had concealed and later bungled the handling of an outbreak of Mad Cow disease (bovine spongiform encephalopathy), at a cost of fifty human lives, four million cattle, and billions of dollars. People began to lose faith in government regulation and monitoring of the food supply. People in Britain questioned their government's ability to protect them from the food they ate. In other countries, including the United States, citizens took a similar position, especially when information—both accurate and inaccurate—about genetically modified food began to appear on the internet, which was growing exponentially in the late 1990s. Increasing numbers of Americans began to demand the labeling of food with genetically modified elements.

In 1997 the cloning of Dolly the sheep was announced. Public opinion polls on cloning in the United States, especially with regard to the prospect of human cloning, consistently reported strong opposition to cloning. Despite long-standing American support for unfettered scientific exploration, investigation, and experimentation—a tradition going back to Benjamin Franklin— Americans remained wary of recombinant DNA and cloning technologies. With regard to cloning and certain types of genetic engineering, for the first time in their history Americans overwhelmingly favored a ban on certain kinds of scientific experimentation, and legislators even advocated the criminalization of certain types of experimentation using certain bioengineering and cloning technology.

As 2001 flowed into 2002, the American economy convulsed under the impact of 9/11 and a wave of corporate scandals. Corporate corruption, cover-ups, fraud, and misrepresentation caused severe popular reservations about the integrity of American corporate culture. That enormous American corporations were marketing genetically modified food, but not informing consumers, did little to restore confidence.

In late 2002, in anticipation of a U.S.-led war against Iraq, a major fissure in European-American relations began to unfold. In the late 1990s, the emergence of a sharp disagreement between the United States and Europe over genetically modified food had already indicated the presence of strained relations. Unlike the war against Iraq, in the conflict over genetically modified food even the British could not be counted on as allies. In 1998, Prince Charles became the British "poster boy" against genetically modified food. Writing in the *Daily Telegraph*, he declared that scientists had strayed into "realms that belong to God and to God alone. . . . We simply do not know the long-term consequences for human health and the wider environment of releasing plants bred in this way [through genetic engineering]. . . . The lesson of B.S.E. [Mad Cow disease], and other entirely manmade disasters on the road to 'cheap food' is surely that it is the unforeseen consequences which present the greatest cause for concern." The owner of "organic" farms, Prince Charles became a leading European spokesperson for "natural foods."

In expressing his opposition to genetically modified foods, Prince Charles articulated three claims often made by the opponents of genetic engineering. The first is that those who engage in genetic engineering usurp divine prerogatives. The second is that genetic engineering is necessarily dangerous and harmful. The third is that the "natural" is always superior to the "artificial." Each of these claims is problematic.

The claim that technology in general and genetic engineering in particular somehow usurp divine prerogatives opposes the theological understanding of the human being as "God's partner in the work of creation." It rejects the biblical claim that the world has been "created to be made," that the human mission includes developing the raw materials created by God and repairing and completing God's creation. Throughout modern history, technological advances, including railroads, airplanes, and refrigeration, have initially been denounced as artificial, as human usurpations of divine prerogatives, as sinful interventions in the natural world. In 1833, when Fredrick Tudor became the first person to use ice to preserve produce on long ship voyages, he was reviled by contemporary theologians as having committed unnatural and sinful acts.

For centuries, farmers have selectively bred plants and animals to produce better food. Farm implements and, later, more sophisticated machines were employed in the process of producing food. Genetic engineering techniques, which aim at refining, improving, and speeding up this process, need not be considered any more of a usurpation of the divine prerogative, or any more of an unnatural act, than the crossbreeding of a horse and a donkey to produce a mule, or the use of tractors or refrigeration to harvest and preserve crops. As is often the case, once new technologies have been demonstrated to be effective, safe, and beneficial to society, they gain public acceptance.

To date, genetically modified food has not been demonstrated to be any less safe than "natural" or "organic" food. Organic food may in fact be less safe than genetically modified food and drugs. For example, the first genetically engineered pharmaceutical, insulin, has been shown to be safer than insulin extracted from the pancreases of cows. The same may be said of much genetically modified food when compared to organic food. Mad Cow disease in England, and periodic massive recalls of meat products in the United States, have nothing to do with genetic engineering. New varieties of vegetables created by traditional "natural" methods of Mendelian genetics have made people sick. Organically grown food fertilized with animal manure has been known to make people deathly ill, as have pesticides—the use of which has been reduced or eliminated in the growing of genetically engineered crops that manufacture their own pesticides.

In many cases, genetically engineered food may be considered as purified and improved food when compared to organic food. Indeed, in the United States, genetically modified food is monitored and tested with much greater scrutiny than "natural" or "organic" food. Furthermore, it is not quite accurate to consider organic food as "natural" and genetically modified food as "artificial." A genetically modified soybean or beet is no less natural, no less edible or real than an organically grown soybean or beet. The sharp dichotomy drawn between the "natural" and the "artificial" is itself often artificial.

Nature is not as benign as "naturalists" would like us to believe. Nature can nurture, but nature can also be cruel and catastrophic, violent and deadly. Nature sustains us. But nature is also the violent explosion of a supernova star, the brutal force of a tidal wave, the devastating impact of a tornado, and the constant struggle of "red tooth and claw" in the animal kingdom. The natural world is a tough neighborhood; it is not the Garden of Eden. Few people today, including naturalists, choose to live "naturally," completely forsaking the benefits of modern technology. Few people really want to return to the state of nature.

Naturalists seem to be asking us to worship nature. But, as we shall see, the opening chapters of Scripture polemicize against the nature worship that characterized the polytheistic religions of the ancient Near East. In modern times, various ideologies, including Nazism, have advocated nature worship and adulation. They readily equate the natural with the good and with the will of God. The laws of nature are invoked as the basis for desirable human behavior, no matter how cruel or destructive. Like the natural world they idealize, such naturalists can be cruel as well as benign. Nazi doctrines of "racial hygiene" invoked theories of "natural selection" to justify their genocidal policies. Explicating the principles of "racial science" for the Reich Ministry of Interior, Dr. Achim Gerke wrote, "Let us not bother with the old and false humanitarian ideas. There is in truth only one human idea, that is: furthering the good and eliminating the bad [extinction of the 'racially inferior']. The will of nature is the will of God. . . .[Nature] sides with the strong, good, victorious. . . . [In exterminating the racially inferior, weak, and impure] we simply fulfill the commandment, no more, no less."

Much of the European aversion to genetically modified organisms, and its advocacy of the "natural," draws upon terrible memories of the implementation of eugenics policies practiced by the Nazis during World War II. What these Europeans fail to realize, however, is the close link between Nazi eugenics policies and the Nazis' emphasis on the "natural."

The growing fissure between European and American culture became increasingly evident during the 2003 war against Iraq, which was strongly opposed by most Western European nations. Fighting to retain their own cultural identities at a time of European Union and America's emergence as the world's only superpower, various European countries perceive genetically modified food to be a symbol of America's alleged desire to impose its technology, culture, and economic power over the lives of Europeans. Because so many European countries have suffered so much, for so long, from foreign invaders, they strongly resist what they perceive to be an American invasion of their cultures and economies. In many European countries where food and cuisine as expressions of national culture are taken very seriously, the "invasion" of American "fast food," "junk food," and genetically modified food is likened to a military invasion.

Europeans carry heavy historical baggage. Because they have known both fascism and famine, they are wary of any attempt to alter the way they live, including the way they farm and the food they eat. For these and other reasons, it should not be surprising that many Europeans are suspicious of genetic engineering, especially when it impacts something as fundamental as food. Simply put, European and American attitudes toward many things—including the nature and preparation of food as well as the "ritual" of the meal—are sharply different and may be irreconcilable. The issue of genetically modified food is a flashpoint of this divergence of attitudes. Food is not simply what we eat. It is also a symbol of cultural, technological, political, and economic values.

Some advocates of genetically modified food see in the new technology a way of addressing the ever-present threat of famine and malnutrition. Although it is not likely that genetically modified food will eliminate these conditions, it cannot be denied that new technologies in farming could be employed to substantially reduce hunger, famine, and vitamin deficiencies on our planet. Restricting the production and distribution of genetically modified food might not deplete but might actually increase human hunger, disease, and famine.

Despite its many advantages, genetic engineering, like any new technology, involves risks. There is a risk, for example, that bioengineered fish and plants might crossbreed with other fish and plants and dominate various species. Despite efforts to segregate genetically engineered species, cross-breeding nonetheless has occurred. There is a risk that insects will grow resistant to the insecticides produced by bioengineered plants and will need to be fought with increasing amounts of new, powerful, and dangerous chemical insecticides. There is a risk that birds and other animals will be harmed by consuming genetically altered plants. As always, there is a risk that new technologies may spin out of control. There is a real and realistic apprehension that once the genetic engineering genie is out of the bottle, its effects on the environment—on plants, animals, and humans—may be irreversible. Although we cannot force already released genies back into the bottle, we can limit the number of new genies that we let loose on the world.

Some observers express legitimate concern over the suffering endured by animals in the production of genetically modified food and pharmaceuticals. Countless numbers of animals, especially pigs and monkeys, live and die painfully in scientific labs in the course of research for the manufacture of new drugs and the development of new therapies for human diseases. Genetic engineering experiments have created grotesque animal chimeras whose short lives are plagued by often harmful and painful physiological abnormalities. Animals created for human consumption, like chickens, live their abbreviated lives in crowded pens where they are force-fed with grain and chemicals. Too cramped to move, too fat to stand, their own feet are often unable to support their abnormal weight. Their short and miserable lives are merely part of the food production process. The use of animals in the manufacture of drugs for human use has been denounced as "pharming."

Animal rights activists have been especially vocal in this regard. They maintain that, unlike genetically altered bacteria and plants that do not suffer and feel no pain, animals are sentient beings. They do suffer and should not therefore be subject to such cruelty. But while the suffering of animals must be avoided whenever possible, the treatment of animals should be seen within the context of what is beneficial for human life and health. Given the choice between genetically altering animals and condemning a group of human beings to suffer from a debilitating disease, we should prefer to mitigate human suffering. Despite cries of "speciesism" from animal rights advocates, animals are not human beings, nor should they be granted the constitutional protections of the Bill of Rights or the ethical protections afforded human beings. As one scientist told me, "No one I know would prefer that we experiment on human babies rather than animals in order to find a cure for a disease that afflicts children."

Advocates of genetically modified food have been characterized by their opponents as demonic creators of "frankenfood." Indeed, the myth of Frankenstein has been invoked by many critics of virtually all forms of biotechnology, including genetically engineered food and pharmaceuticals as well as reproductive biotechnology. But the real Frankenstein monster in genetic engineering is not technology. Rather, it is corporate manipulation, greed, and deception.

Genetically modified foods were largely imposed upon American consumers without their knowledge or assent. Although many farmers readily acknowledge the many benefits afforded by the new farming technologies, they also are increasingly aware of their growing control by the corporate food industry. For example, bioengineered seeds can be acquired only under agreements that restrict what a farmer can grow—that is, how the seeds may be used. These seeds cannot be used from season to season but must be purchased each year from a corporate seed manufacturer and distributor. Since seeds are patented, their use is tightly restricted, with stiff legal penalties for noncompliance with policies determining their use. As the number of life-science companies has shrunk through corporate mergers, fewer and fewer companies are controlling more and more farmers and what they produce.

Huge multinational corporations are attempting to impose their agendas and fiscal goals not only on individual farmers but on governments as well. The food chain is coming under the control of a small number of corporate entities. The ability to control food is a powerful and potentially dangerous weapon in the hands of corporate officers driven primarily by the balance sheet. And corporate giants in effect pursue their own foreign policy, which may be at odds with that of their country; or they try to manipulate foreign policy to their own economic advantage. The strong-arm tactics of multinational (largely American-based) food corporations have alienated many European governments and exacerbated foreign fears about American attempts to influence their national policies. Economically developing countries are especially wary of succumbing to the economic and political influence of American-based multinational biotechnology corporations. The "Big Brother" that greater numbers of people worry about is not governmental but corporate. Corporate power, greed, governmental influence, patent protections, and legal resources have become too formidable for most to resist.

Had Rabbi Hanina and Rabbi Oshaya lived today, their lawyers would have encouraged them to apply for a patent on the golemic calf they created through mystical means. By extending U.S. patent law to genetically engineered organisms, the Supreme Court has not only helped stimulate the exponential growth of the biotech industry but has also established a precedent for human beings and corporations to enjoy monopolistic commercial rights of ownership over entities that are now part of the natural world. The precedent established by one vote on the Supreme Court in *Diamond v. Chakrabarty* was extended in 1987 by the Trademark and Patent Office to include other forms of life, including plants and animals. Congressional legislation now permits universities to be granted patents directly, to license patent rights, and even to collect licensing fees for products developed with federal research funds.

As a consequence, a university-industrial complex of patent-controlling academic institutions and commercial corporations has evolved. Today seed, plants, animals, bioengineered pharmaceuticals, human tissue, human blood, and human genes, as well as processes for producing and testing them, are owned or licensed by a small circle of universities and multinational commercial corporations. Although parts of the human genome and the human body, and many methods to determine the status of its health, are corporately owned or licensed, only the Thirteenth Amendment—which prohibits the ownership of a human being—stands in the way of patenting the whole human person. Patent rights are vigorously defended by the corporate giants that hold them, and a routine patent infringement case that goes to trial generates more than $1 million in legal fees, which only large corporations can afford.

During the summer of 2001, one of the first challenges faced by the new Bush administration had to do with the use of embryonic stem cells for biomedical research. Only weeks after the president announced his administration's position on stem-cell technologies and cloning, a greater challenge confronted him: the 9/11 terrorist attack on American soil. In the months that followed, public anxiety about the government's ability to protect its own citizens was pervasive. The fear now was not genetically engineered foods but bioengineered chemical and biological weapons. Again, bioengineering was front-page news.

While a great many applications of genetic engineering may be considered ways of fulfilling the biblical mandate to improve and complete nature, the creation of new and improved weapons of mass destruction through bioengineering is not among them. It is the purpose and not the process that determines the ethical propriety of developing and employing the methods and products of bioengineering. Harming and murdering people cannot be considered a viable purpose of the new technology.

With advances in science in the latter half of the twentieth century, deadly viruses joined other biological and chemical weapons in the arsenal of weapons of mass destruction. New delivery systems were developed to help them reach intended targets. Such weapons could be as mildly disabling as the flu or as deadly as an atomic bomb. In 1969, President Richard Nixon began a process to initiate a worldwide ban on chemical and biological weapons, and to focus American research on defensive measures in combating such weapons. In Nixon's words, "The human race already carries in its hands too many seeds of its own destruction." Or, as the novelist and physician Michael Crichton has put it, "Genetic engineering can do more damage than nuclear bombs."

Recombinant DNA technology has the potential to create weapons of mass destruction that dwarf all those that came before in their power to harm and destroy human, animal, and plant life. As this technology has developed, it has become easier and cheaper to employ. Unlike the atomic or hydrogen bomb, it does not require huge sums of money to produce. Unlike atomic weapons, its presence is often difficult to detect. Indeed, such weapons of mass destruction could readily become "a poor man's atomic bomb." One observer has called them "a low-rent way to be a big player." Such comparatively cheap weapons of mass destruction have helped to level the playing field in international and intergroup conflict. As we saw in 2001, even a small amount of anthrax sent through the mail could provoke a major disruption in American life. If such a relatively unsophisticated small-scale attack could cause such anxiety and damage, imagine what the unleashing of an arsenal of biological and chemical weapons against our cities or our food or water supply could accomplish. Despite efforts to take defensive measures against such attacks and to seek antidotes to their potential effects, offense is usually ahead of defense. Many of the drugs recently designed to combat dangerous bacteria are no longer effective because the bacteria mutate faster than our ability to devise even more potent drugs to combat them.

The ultimate use of genetic engineering is its application to human beings. It is one thing to try to genetically modify plants and animals; it is quite something else to consider genetically altering our own selves. With genetic engineering, human beings have taken evolution into their own hands, including human evolution.

The eminent physicist Freeman Dyson tells of the time his five-year-old first saw him naked. "Did God really make you like that?" the child asked. "Couldn't God have made you better?" she continued. To which Dyson responded, "Yes, of course God could." But God didn't. That's our job. Our mission according to religious traditions such as Judaism, is to improve the world and improve ourselves, not only spiritually but physically as well.

We readily use technology to try to improve upon what nature has dealt us, from the eyeglasses we wear to improve our vision, to the pharmaceuticals we ingest to fight and control disease, to the surgeries we sometimes must undergo to save and enhance our lives. With the publication of the human genome and the inception of genetic engineering, still more possibilities have become available. Should we take advantage of them, and if so, how far ought we to go?

In his groundbreaking book *Remaking Eden* (1996), Lee Silver, a biologist at Princeton University, tells how he asked his students whether they would consider using genetic engineering on their future children for any reason whatsoever. The response was generally that they would not. But when Silver then asked them whether they would use genetic engineering if it could provide absolute protection against a disease like AIDS, most changed their minds. This informal classroom survey echoes the attitudes of most Americans toward the application of genetic engineering to human beings. We tend to oppose it on a gut level but to embrace it if we perceive it to be purposeful and beneficial.

Genetic engineering has the potential to treat a wide variety of human diseases and to enhance our physical and intellectual capabilities. For example, therapies are now being developed to alter genes to prevent children from inheriting diseases such as diabetes. Genetic therapies are under way to replace bypass surgeries by growing new arteries to replace damaged ones. Other planned genetic therapies aim at fighting cancer by preventing tumor-suppressing genes from "shutting off." The potential use of "artificial chromosomes" (developed in 1997) to add various features to our genetic composition is also under study. Genetic "enhancement therapies" to make us smarter, stronger, and even better-looking are being devised.

Some of these therapies aim at curing or preventing genetic defects and addressing genetic predispositions to certain diseases, thereby eliminating or lessening the eventual need for invasive surgeries or drugs. Others aim at improving on the hand that nature has dealt us by enhancing our current capabilities. Some of these will be somatic gene therapies that will alter the genes in our body but will not be passed on to future generations. Others will be interventions at the embryonic state that will not only alter the genes of the child to be born but will become incorporated into the child's genome and passed on to succeeding generations. In this way, many deadly hereditary diseases may be prevented in future generations. Still other therapies envision using "artificial chromosomes" that might be replaceable once new and better "models" became available. For some observers, the development and use of advanced genetic therapies are both inevitable and desirable. For others they are neither. The basic issue here is what kinds of human beings we want ourselves and our children to become, and at what cost.

A May 11, 2001, article in the *New York Times*, "Someday Soon, Athletic Edge May Be from Altered Genes," reported that a 1995 survey of two hundred aspiring American Olympians asked whether they would agree to take a banned drug that would alter their genes in a way that would guarantee victory in every competition for five years—but would soon after cause their death. Over half of these young athletes said yes. This and similar surveys indicate that we tend to focus on short-term benefits rather than long-term consequences. We are not always wise enough to discern when the employment of new technologies is beneficial in the long run.

As is often the case, science fiction articulates popular fears about the implications of new technologies, including the application of genetic engineering to human beings. For example, the 1982 film *Blade Runner* portrays genetically engineered "replicants" who are superior to other humans who do not avail themselves of genetic therapies. The trade-off is that the replicants collapse and die after only four years. Like the athletes discussed in the *New York Times*, they enjoy short-term gain and suffer premature death.

The 1997 film *Gattaca* envisions a society in which a genetically engineered upper class oppresses those who have not been chosen for genetic altering. *Gattaca* alludes to the four proteins of the human genome: GATC. Like the biologist Lee Silver, this film anticipates a future ruled by the "Genrich" (the genetically enriched), constituting about 10 percent of the population, and a subservient underclass of human "naturals" that includes the rest of society. Yet all of this lies in a future that may never come. At present, the genetic engineering of human beings remains in a nascent stage.

In January 1997, in his second inaugural address, President Bill Clinton remarked, "Scientists are now decoding the blueprint of human life. Cures for our most feared illnesses seem close at hand." The statement was naive and misleading. Although we have a text of the human genome, we have yet to unravel its meaning. The process of understanding and interpreting the text is still in its infancy. We have yet to learn what genes do, how they work, what proteins they make, how they interact. Although greater numbers of genetic screening tests are now available, there is a wide gap between the diagnostic and the therapeutic.

We can identify certain genetic predispositions toward a variety of diseases, but we cannot yet treat most of them. While genetic screening for disease has been improved, we often do not know what to do with the information it provides. For instance, genetic screening can reveal a predisposition to a disease that is currently incurable and untreatable. Despite legal restrictions on genetic discrimination, people who undergo genetic screening suddenly find themselves uninsurable, unemployable, and unsuitable for certain prospective marriage partners. Since most of us have eight to twelve genetic defects, the more we screen the more problems we will face in treating the information we acquire.

Although a genetic predisposition to a disease does not mean that we will fall victim to that disease, the information collected by genetic screening compels us to rethink our self-image and reassess our future. Without doubt, such information would influence our way of living, our self-image, our future planning, and how others relate to us. As the legal scholar Lori Andrews has observed, "People's most intimate sense of themselves and feelings of security can be shaken by the use of genetic services."

But there is still another side to the situation. One of the success stories of genetic screening relates to Tay-Sachs, a disease disproportionately found among Ashkenazic Jews. The offering of free testing and counseling has caused this disease to be virtually eliminated in many segments of the Jewish community. But had the test for detecting Tay-Sachs been restricted, the patent holder could have shut down genetic testing altogether or have made it prohibitively expensive.

In its code of ethics, the American Medical Association has prohibited its members from patenting procedures because it has found that such patents compromise the quality of medical care. The medical organization of genetic specialties, the American College of Medical Genetics, also opposes gene patenting. Meanwhile biotech companies patent genetic tests, genetic therapies, and human genes, thereby potentially compromising medical care and driving up medical costs to the point where they exclude many people from access to care, even when some of the research utilized has been financed by public funds—our tax dollars.

Various forms of genetic therapy hold great promise for the treatment of many diseases. At the time of this writing, they have an uneven record of success. For example, genetic engineering has been successfully used after bypass surgery to prevent blockages that occur in the coronary arteries and the arteries of the legs. On the other hand, the September 1999 death of Jesse Gelsinger dealt the entire field of genetic therapy a severe blow. Gelsinger had undergone gene therapy for an inherited enzyme deficiency. He apparently suffered a severe immune reaction and died four days after being injected with the engineered virus. The researchers did not fully understand what went wrong.

Reflecting on the attitude of Jewish tradition toward genetic screening and therapy, the prominent physician and bioethicist Fred Rosner writes, "Genetic engineering, gene therapy, and other applications of genetic engineering are permissible in Judaism when used for the treatment, cure, and prevention of disease. Such genetic manipulation is not considered to be a violation of God's natural law but a legitimate implementation of the biblical mandate to heal. According to Jewish law, if Tay-Sachs disease, diabetes, hemophilia, cystic fibrosis, Huntington's disease or other genetic diseases could be cured or prevented by gene surgery, it is certainly permitted. . . . The main purposes of gene therapy are to cure diseases, restore health, and prolong life, all goals within the physician's Divine license to heal. Gene grafting is no different than an organ graft such as a kidney or corneal transplant, which nearly all rabbis consider permissible."

Rosner is, of course, speaking mostly of gene therapies not yet available; or, if available, not yet considered adequately developed for public application. Rosner also considers many genetic-enhancement therapies—since they have no clear therapeutic purpose—to be frivolous. But the line between therapy and enhancement is not always clear. Who will determine distinctions between them—physicians, insurance companies, patients, bioethicists?

For many observers, genetic therapies aimed at curing and treating existing diseases and disabilities are desirable, but genetic therapies aimed at enhancing certain conditions that are neither life nor health threatening are frivolous and even dangerous. Such enhancement therapies serve to "raise the bar" for what is "normal" and desirable, setting a standard that will remain out of reach for many who find it desirable. Many examples might be given. For instance, when growth hormones became available, they were widely used to help ensure that short children would be able to attain "acceptable" height. Parents did not want their children to be stigmatized or discriminated against for being "short." Thus growth hormones that were once restricted to combat dwarfism were given to thousands of children who were not threatened with dwarfism. The widespread use of these hormones, especially in view of their possible side effects, went unchecked.

That parents would want enhancement for their children is understandable. Many parents want more intelligent children—but heightened intelligence has been linked to aggressive and even anti-social behavior. Do we wish to extend genetic enhancement to culturally imposed views of beauty or desirability, like hair and eye color, and sex selection? Prenatal sex selection has been used to prevent the passing down of certain genetic diseases, such as hemophilia, from mothers to sons. But to what extent do we use prenatal sex selection, and for what reasons? Will the genetic screening of fetuses lead to "frivolous" abortions, as right-to-life advocates fear? Consider, for example, that in one survey 12 percent of potential parents said they would abort a fetus with a genetic predisposition to obesity. What of a predisposition to certain diseases and disabilities? Certainly parents want to provide their children with every advantage, but where should we draw the line? How far should we go?

With regard to the possible use of genetic engineering on our children, the bioethicist Arthur Caplan has written, "We mold and shape our children according to environmental factors. We give them piano lessons and every other type of lesson imaginable. I'm not sure there is anything wrong with using genetics . . . as long as it is not hurting anyone or . . . ideas of perfection [are not being imposed] on anybody." But legal scholars are already asking whether parents who do not have their children screened for certain genetic disorders can be held liable if such disorders surface later in life. Should children with genetic disorders, or children who do not believe their parents gave them adequate "enhancement" therapies, such as growth hormones, be able to sue their parents for child abuse?

Already categories that echo twentieth-century eugenics policies, such as "wrongful life," are entering our legal lexicon. The specter of Nazi policies hangs over our deliberations about whether to apply genetic technologies to human beings. Like all technologies, eugenics can be benign or malignant, beneficial or genocidal. The determining factor is whether it is freely chosen by individuals or imposed by state or corporate power or by the pressures of social compliance.

Historically it would be a mistake to think that malevolent eugenics policies and their application can occur only in a criminal regime like Nazi Germany, that they cannot happen in a country like the United States. Indeed, they already have.

The American eugenics movement was spawned in the 1890s as massive waves of immigrants arrived on American shores. Prominent American statesmen, jurists, and academicians, like Theodore Roosevelt, Justice Oliver Wendell Holmes, and Professor Charles Davenport of the University of Chicago, were among the leading American advocates of eugenics policies. So were early feminists like Margaret Sanger. One of the reasons Sanger was an early advocate of birth-control programs is that she wanted to limit the number of society's "undesirables." In Sanger's words, "It is a curious but neglected fact that the very types which in all kindness should be eliminated from the human stock, have been permitted to reproduce themselves and to perpetuate their group. . . . [Government imposed sterilization is the solution to] take the burden of the insane and the feebleminded from your back."

Or, as Theodore Roosevelt put it, "Some day we will realize that the prime duty, the inescapable duty of the citizen of the right type is to leave his blood behind him in the world; and that we have no business to permit the perpetuation of citizens of the wrong type. The great problem of civilization is to secure a relative increase of the valuable as compared with the less valuable or noxious elements of the population. . . . I wish very much that the wrong people could be prevented entirely from breeding; and when the evil nature of these people is sufficiently flagrant, this should be done. . . . The emphasis should be laid on getting desirable people to breed."

Only with the need to enlist new immigrants into the American armed forces during World War I did the eugenics movement begin to fade in the United States. But its early advocates in Nazi Germany found a precedent for their policies in the American eugenics movement and in American legal opinions advocating the compulsive sterilization of criminals and "imbeciles," such as those promulgated by no less of an icon of American jurisprudence than Oliver Wendell Holmes.

"What we racial hygienists promote," said a leading advocate of eugenics at the time in Germany, "is not at all new or unheard of. In a cultural nation of the first order, The United States of America, that which we strive for was introduced long ago. It is all so clear and simple." Later in Nazi Germany, eugenics deteriorated into genocide.

Like the ancient and medieval rabbis who created golems, we are confronted with the question of whether to take the risk of creating new life-forms and re-creating existing ones, including ourselves. The challenge is whether or not to use the knowledge and technological abilities that we now have or soon will. On the one hand are dangers and uncertainties that advise us to desist. On the other hand are benefits and blessings that prompt us to continue. Imposing new technologies can alter our lives, our world, and ourselves in horrific ways, but withholding them can deprive us of enhancing and improving the world, the human body, and the life provided us by nature.

As he walked with his disciples to the river in Prague to create his golem, questions such as these undoubtedly percolated in the mind of Rabbi Judah Loew. He created his golem nonetheless. And, as rabbinic legend tells us, when God contemplated creating the first human being, God had similar misgivings. Consulting with the heavenly host, God was advised not to proceed with the creation of human beings because it was too risky to bring such a dangerous creature into the world. Nonetheless God created human beings. Although the human creature did not turn out the way its creator expected, in all his omniscience and omnipotence God created us in any case, and God "created a world to be made" by us—for better or for worse.

The challenge posed by genetic engineering has been concisely articulated by Rabbi Barry Freundel, who served as a consultant to the Presidential Commission on Cloning. He said, "I do not find [human beings gaining control of their own evolution] to be any more troubling than discussing any other human capacity to alter the natural world. . . . God has entrusted this world to humankind's hands, and the destiny of this world has always been our responsibility and our challenge. Whether or not we live up to that challenge is our calling and essential mission. . . . If God has built the capacity for gene redesign into nature, then He chose for it to be available to us, and our test remains whether we will use that power wisely or poorly."

Chapter Seven

Test-tube Golems: Stem Cells and the Cloned Arranger

"Sometimes the thought that she would never give birth tore her heart. . . . Puttermesser asked: 'When you woke into life, what did you feel?' 'I felt like an embryo,' the golem wrote." This is a slice of dialogue between creator and golem in Cynthia Ozick's novel *The Puttermesser Papers*. Puttermesser, an aging civil servant in the New York City bureaucracy, feels the gnawing anguish of being childless. In her desperation and to her surprise, she successfully creates a female golem to be her daughter.

Throughout human history, countless women and men have felt the sting of being childless. The biblical matriarch Rachel prays for death rather than endure barrenness. In recent decades, especially in highly developed countries, human infertility has risen to epidemic proportions. It has been estimated that as many as 1.2 million patients in the United States are treated annually for infertility, that more than one of every six married American couples suffers from infertility. In response to the growing need for medical intervention, new methods of reproductive technologies have been developed and applied.

In the highly regulated American health-care industry, a governmentally unregulated subindustry dealing with human infertility has mushroomed. Procedures that were once considered medically dangerous and morally heinous have become routine, including donor insemination, egg donations, and in vitro fertilization (IVF). What was once the fantasy of science fiction has become standard practice in many infertility clinics throughout the world, with even more daring innovations predicted yet to come. Some experts foresee the use of artificial wombs, such as those described by Aldous Huxley in his modern classic, *Brave New World.* Others see a not too distant future in which the reproductive cloning of human beings will simply be one of a variety of routinely available methods of bringing children into the world.

Many techniques now used in human reproductive biotechnology were first employed with animals. The in vitro fertilization of a rabbit was first accomplished in 1954, and of a human in 1978. In 1972 a mouse was born from an embryo that had been frozen, in 1973 a calf, and in 1984 a human being. Donor insemination, embryo transfers, and "surrogate mothers," long used in animal husbandry, are currently being used in the treatment of human infertility.

The sale and purchase of human sperm and eggs have become big business. Pick up the average college newspaper and you will find ads offering and soliciting human sperm and eggs. One such ad, presumably by a student looking for a way of helping to pay her substantial tuition, reads: "Jewish egg donor. . . . Very attractive, in perfect health. Willing donor for couple seeking great genes." Eggs and sperm are available for sale even on the internet. There one man advertised a vial of his sperm for $4,000, assuring potential buyers that he could trace his lineage back to six Catholic saints and several European royals. Fashion models sell their eggs through cyberspace at prices ranging from $15,000 to $150,000. Donated sperm may be used for artificial insemination and for in vitro fertilization. Donated eggs may be used for implantation either before or after fertilization, in vivo (in the body) or in vitro. Frozen IVF embryos may be preserved for future implantation.

As early as 1934, Herman Rohleder wrote a history of the artificial impregnation of human beings—that is, through artificial insemination. That work was entitled *Test Tube Babies.* On July 25, 1978, Louise Brown, the first child conceived in vitro, the first "test-tube baby," was born at a hospital in England. The sperm of her father, John Brown, had fertilized the egg of her mother, Lesley Brown, in a Petri dish. Three days later Dr. Robert Edwards implanted the embryo into Lesley's womb. Under the surveillance of physicians and an aggressive press, the pregnancy progressed, and—to the surprise of many—a healthy baby was born. Today Louise Brown lives in England. Since her birth, hundreds of thousands of children have been conceived through in vitro fertilization. While the success rate of this method is about 16 percent, it is estimated that by 2005 about half a million people will have been conceived in this manner in the United States.

The success of Dr. Edwards, an embryologist, and his obstetrician collaborator, Dr. Patrick Steptoe, was met with an avalanche of criticism and abuse. Earlier, in 1971, James Watson, the co-discoverer of DNA, had chastised Edwards: "You can only go ahead with your work if you accept the necessity of infanticide.

There are going to be a lot of mistakes." Scientists criticized Edwards for not having first experimented on higher animals. The church and the British Parliament accused him of immoral deeds, of unnatural acts, of "playing God." Various states in the United States immediately banned IVF after the birth of Louise Brown. In the late 1970s in Illinois, in vitro was deemed illegal on the grounds that it was a form of *child abuse*. Similar views are now being expressed by the opponents of reproductive cloning.

In 1966 the physicians Sophia Kleegman and Sherwin Kaufman, in their book *Infertility in Women*, wrote, "Any change in custom or practice in this emotionally charged area [of assisted reproduction] has always elicited a response from established custom and law of horrified negation at first; then negation without horror; then slow and gradual curiosity, study, evaluation, and then a very slow steady acceptance." Despite initial strong opposition to in vitro fertilization, recent polls indicate that more than 75 percent of Americans now believe it to be an acceptable solution to human infertility.

Throughout the world in 1978, media reports both praised and vilified the birth of the first test-tube baby, the first human being conceived outside the womb. In its reporting of this event, *Time* compared the birth of Louise Brown to the creation of a golem in that both had been brought to life through "artificial" means. In an obscure Hebrew journal on contemporary issues in Jewish law, Rabbi Judah Gershuni—in the course of discussing when and how this method of conception might be permitted according to Jewish law—compared in vitro fertilization to the creation of a golem.

Since the birth of Louise Brown, many new methods for treating human infertility and improving the potential for conception, have been developed. Since 1992, ICSI (intra-cytoplasmic

97

sperm injection) has been used in cases where sperm quality is extremely poor. In this procedure an individual sperm is physically forced through the membrane of the egg to effect fertilization. This method can obviate the use of a sperm donor in situations where a married couple desires a child that will have their own genes rather than those of a third party. ICSI relies upon IVF. The sperm and egg are joined in vitro, and the embryo is then implanted for gestation. IVF also has made possible another form of reproductive biotechnology, cloning. Many infertility clinics currently have the technology to do ICSI and IVF; therefore technically they also have the ability to do SCNT (somatic cell nuclear transfer), which is a crucial step in reproductive cloning.

IVF is not risk free. The technique begins by injecting a woman with hormones to stimulate ovulation. This procedure slightly increases risks of stroke, heart attack, and ovarian cancer. It may also offer some risks to the embryo and eventually to the child. Although American fertility clinics do not keep records of genetic abnormalities in children born as the result of IVF and/or ICSI, clinics in other countries such as Australia do. Some data indicate a higher incidence of congenital defects among children brought into the world in this way when they are compared with the general population.

As early as the 1950s, cloning was accomplished using amphibians such as frogs. At the time many scientists doubted that it could be done with mammals. In cloning, the nucleus of an egg cell is destroyed, and the nucleic matter of a cell from a "donor" is infused into the egg cell. The new cell is then artificially stimulated (by chemicals or electricity) to divide in vitro. At the appropriate stage it is implanted in a womb, where hopefully it gestates to the live birth of a healthy offspring.

Mammalian cloning was discussed in scientific journals throughout the 1950s and 1960s. Most scientists were convinced that mammalian cloning, especially from adult donor cells, was extremely unlikely, and that human cloning was a scientific impossibility. The Nobel laureate Joshua Lederberg disagreed.

In 1966, Lederberg predicted that human cloning was a matter not of "if" but of "when." In response, in 1970, the eminent theologian and bioethicist Paul Ramsey unequivocally denounced human cloning as immoral. He was followed by others who rejected cloning as immoral, unnatural, repugnant, and as "playing God." Human cloning, many claimed, could not be compared to other developments in reproductive biotechnology but must be viewed as posing new and dangerous challenges to our conceptions of human nature, human identity, human reproduction, and human parenthood.

In 1972 the psychiatrist Willard Gaylin, in an article in the *New York Times Magazine*, catapulted the discussion of cloning from the realm of science and academe into the public arena. Entitled "The Frankenstein Myth Becomes a Reality: We Have the Awful Knowledge to Make Exact Copies of Human Beings," the article was illustrated with multiple pictures of Mozart and Hitler. Although Steptoe's and Edwards's first in vitro birth was then still six years in the future, Gaylin noted their research in IVF research as a critical step in the process that would eventually lead to human cloning.

Still, in the years after Gaylin's article appeared, it seemed a moot issue, more a matter of science fiction than public policy. Then, in February 1997, when the birth in July 1996 in Scotland of a cloned sheep named Dolly was publicly announced, cloning became a topic of intense public discussion. In the cloning of Dolly, scientists had succeeded in resuscitating a mature cell that had reached its final differentiation by returning its embryonic characteristic of developing into a complete organism. Although many scientists had rejected the possibility that a mammal could be cloned from an adult cell (in the case of Dolly, from a cell taken from the frozen udder of a dead sheep), the birth of Dolly proved that it was indeed possible.

Immediately President Clinton and British Prime Minister Tony Blair were joined by a chorus of legislators, theologians, bioethicists, and (according to surveys) the vast majority of the American public in insisting on a ban on any attempts to clone a human being. According to a 2001 TIME/CNN poll of Americans, 90 percent believed it a bad idea to clone a human being, with only 7 percent favoring it. To the question, Is it against God's will to clone humans?, 69 percent said yes and 23 percent no. In 1998 the United Nations endorsed the view that "Practices which are contrary to human dignity, such as reproductive cloning of human beings, should not be permitted." In recent years, bills introduced in the U.S. Congress have called for jail terms of up to ten years and for fines of up to $1 million for attempts at any type of human cloning.

No issue of biotechnology has initially evoked a stronger public outcry than the possibility of reproductive human cloning. In the words of Leon Kass, chairman of the President's (George W. Bush's) Council on Bioethics, "The prospect of human cloning, so repulsive to contemplate, is the occasion for deciding whether we shall be slaves of unregulated progress, and ultimately its artifacts, or whether we shall remain free human beings who guide our technique toward the enhancement of human dignity."

In late 1998 a new scientific discovery was unveiled that renewed public discussion about cloning. Scientists announced the isolation of human embryonic stem cells. These were "pluripotent" cells derived from human embryos—cells that could potentially produce all types of differentiated cells, including heart, liver, pancreatic, and nerve cells. Scientists believed that the use of these cells held great promise for curing many diseases. But in order to isolate these embryonic stem cells, produced through IVF, the embryos that contained them would have to be destroyed in the process. Yet once these cells were isolated, they could be continuously replicated for research aimed at addressing a variety of medical conditions.

The first major social policy issue that confronted the presidency of George W. Bush in the summer of 2001 was stem cell research, particularly the use of government funds to support biomedical research with embryonic stem cells. Bush agreed to the use of existing embryonic stem cell lines for research but opposed the use of newly created embryonic stem cells or of available embryos that had been created for infertility patients through IVF. These restrictions on "therapeutic" human cloning research were coupled with an absolute prohibition against "reproductive" cloning under any circumstances in the United States. At the heart of the matter was the issue of the status of a human embryo. As a strong pro-life, anti-abortion advocate, Bush believed that human embryos—at any stage of gestation—must be considered human beings, with all the rights and legal protections of that status.

Bush appointed the Kass commission to study the medical and moral implications of human cloning, and to submit recommendations to him. In July 2002 the commission unanimously recommended a ban on all human reproductive cloning, calling it an intrinsically unethical act. On therapeutic human cloning, or "cloning for biomedical research," a majority of the commission recommended a four-year moratorium, and a minority—"eager to see the research proceed"—cautiously recommended that it be permitted, though under strict federal regulation. Meanwhile the public policy debate continued in the halls of Congress, in the media, and in the public square.

The therapeutic potential of stem-cell research altered the focus of the public discussion of cloning. It persuaded some of the staunchest opponents of cloning and abortion to reconsider their position. Although opposition to reproductive cloning—on philosophical, theological, and medical grounds—remained strong, the use of stem cells, including embryonic stem cells, in biomedical research gained many supporters.

Stem cells hold the promise of substantially broadening the scope and nature of medical therapies to include what is often called "regenerative medicine." This approach has the potential to restore lost organ function and to repair organs severely impaired by disease or injury—currently impossible. The strategy of this new approach aims at repairing, rather than simply halting, the damage done to various organs by injury or disease.

Stem-cell therapies may cure or eliminate many diseases and medical conditions, including juvenile diabetes, Parkinson's disease, Alzheimer's disease, stroke, various types of heart disease, and spinal cord injuries. Regenerative therapies could be applied to a wide variety of diseases—repairing damaged tissue of the heart, for example. Organs beyond surgical or pharmaceutical treatment might be regenerated.

In the treatment of diabetes, stem cells could be stimulated to become islet cells, which could then be injected into the pancreases of insulin-dependent diabetics. These cells could then duplicate themselves, allowing the pancreas to produce its own insulin. This procedure would not only "cure" diabetics and eliminate their need for insulin but would also prevent the secondary consequences of diabetes, such as blindness, nerve damage, kidney damage, skin ulcers, and circulatory complications.

Stem cells that change into nerves could replace neuronal tissue damaged by strokes, spinal-cord injuries, ALS, Alzheimer's, and Parkinson's diseases. Skin tissue grown from stem cells could provide a ready supply for severe burn victims. Lab-grown cardiac tissue could shore up damaged arteries and hearts. Injecting stem cells into a liver could produce new liver cells, regenerating an organ damaged by cirrhosis or hepatitis. By replacing worn cartilage, stem-cell therapies could address rheumatoid arthritis and osteoarthritis. Stem cells that morph to produce cardiac muscle could replace cells damaged by chronic heart disease. Through embryonic cloning, scientists could also learn more about the etiology of a wide variety of diseases.

Stem cells may be taken from adults or from embryos as well as from umbilical cords. Today expectant mothers are being asked whether they want the placenta preserved after birth for future use in stem-cell therapies. Although there have been some remarkable breakthroughs with the use of adult stem cells, they are not considered as promising as those from embryos. Adult stem cells are believed to be multipotent—able to change into a number of different kinds of cells—whereas embryonic stem cells are pluripotent—potentially able to become any kind of cell. Adult stem cells seem to have more limited use than embryonic stem cells. Most scientists therefore believe that embryonic stem cells have the greatest potential application for regenerative medicine, and that the best strategy would be a research program in which both embryonic and adult stem cells are used. In this way their potentials could be compared and they could each be used in the most effective way.

Not only does the application of regenerative medicine promise to "cure" diseases that now can only be "treated," it also promises to largely eliminate the need for invasive surgeries and toxic pharmaceutical therapies. Each of us may be harboring the seeds of our own self-regeneration and self-renewal. Stem cells may be those seeds. It has been estimated that 100 million Americans who suffer from a wide variety of diseases and conditions might be helped by this therapeutic approach.

To date, remarkable progress and sound scientific data indicating the potential success of regenerative medicine is readily available. Yet substantial research and experimentation remains to be done if we are to learn the full potential of this new approach to medical care. Just as the potential harm from cloning technology remains unknown, so do its potential benefits. What we do know is that banning cloning research—especially with embryonic stem cells—could foreclose discoveries and therapies with substantial benefits. We also know that the potential risks of cloning technology are not likely to be eliminated unless experimentation and research proceed.

Various biotech companies are now trying to discover a way to retard aging and increase longevity through the use of cloning technologies with stem cells. Yet philosophers and theologians have condemned such attempts as unnatural and as "playing God." Leon Kass has declared, "The finitude of human life is a blessing for every individual, whether he knows it or not." At a conference held in Philadelphia in 2000 on the employment of these technologies, their only advocates were the scientists developing them and a single Jewish theologian. The theologian correctly pointed out that not only is the extension of life obligatory, but that if life is at stake, even certain strictures of Jewish law must be violated in order to extend it. To the question of whether therapeutic cloning aimed at sustaining and extending life is "playing God," one physician commented, "We don't want to play God, we only want to play doctor."

After Dolly the sheep was born in 1996, the vast majority of religious authorities in America condemned human reproductive cloning (the potential of stem cells had not yet been discovered) and asked for it to be banned on moral grounds. According to one Christian evangelical theologian, "Cloning a human baby isn't just bad, or unfortunate, but something which would be profoundly evil because it would constitute a new human being in a radically defiled and deformed moral fashion." To the surprise of many, spokespersons for one religious group dissented— traditionally religious Jews.

Unlike most other religious bioethicists, Jewish bioethicists did not oppose reproductive cloning in principle. Writing and speaking about it from the perspective of classical Jewish law and ethics, leading Jewish bioethicists concluded that cloning is not necessarily prohibited by Jewish religious tradition, that it is neither unnatural, morally reprehensible, nor theologically objectionable. As Dr. Abraham Steinberg of Hadassah Medical School in Jerusalem put it, "The technology of human cloning poses no threat to the belief in the Creator of the world and the Creator of humankind. There is not, and should not be, any shifting of basic values or beliefs in light of these technological advancements." Steinberg explained, "Cloning differs only in technical method in the creation of a fetus. With regard to cloning, we are speaking of the creation of humans 'extant to extant' which diverges from nature in technique alone and not in substance. Similarly, these technologies do not solve the mystery of life, disclose life's basic essence, or claim to create something from nothing which is God's purview alone."

Jewish bioethicists did not endorse cloning carte blanche. They agreed that it should be permitted but regulated. For example, they agreed with secular legal scholars that cloning would be both illegal and immoral if it were done against a person's will. They insisted that cloning be deemed safe both to mother and child before it was done. Although they did not endorse indiscriminate reproductive cloning, they identified situations in which cloning could be justified—for example, an infertile couple with no other chance of having a child. One example given in testimony before Congress by a Jewish bioethicist was the case of a Holocaust survivor whose children had been murdered by the Nazis, for whom cloning would be the only way to have a child.

If animals in danger of extinction are cloned to preserve their species, would opponents of reproductive cloning oppose it if the human species were so endangered? If each human person represents the entire human species, as the Talmud observes, should not a person with no other reproductive option be permitted to be cloned? For couples who want a child with some genetic link to at least one of their parents, cloning may be the only option. Women who have prematurely experienced menopause; women who have undergone a hysterectomy in a previous marriage; men who have been castrated for medical reasons; gay men and women—all might look to cloning rather than to adoption or egg donors or sperm donors. Because of the costs and potential dangers of reproductive cloning, it is not very likely that it would become a pervasive practice, especially in the near future.

Although Jewish bioethicists do not consider human reproductive cloning as morally reprehensible in principle, all insist that safety should be a major consideration. As long as there are overt dangers to the mother or to the cloned embryos, cloning should not be practiced. This includes the risk to the cloned individual of premature death, genetic disease, and the onset of other diseases and maladies due to the employment of the cloning technique. In many animals that have been cloned, such conditions have appeared, including extreme obesity, genetic defects, and premature aging. But, according to Jewish bioethicists, once these dangers have been reasonably addressed, the reproductive cloning of human beings could proceed with caution, under strict regulation, and for specific justified purposes. We can assume that scientists will improve the technology of reproductive cloning, allowing for a greater percentage of healthy animals and humans.

All reproduction involves risk. When the risks of reproduction through cloning approximate the risks of reproduction through coitus, or through currently accepted methods of reproductive biotechnology, the major impediment to permitting human cloning will be eliminated, according to Jewish law. Some scientists believe that cloning may eventually become safer than sexual reproduction because it bypasses the most common form of birth defect—the child's having the wrong number of chromosomes, which can cause various genetic diseases. (As women age, the risk of having too many or too few chromosomes rises.) Reproductive cloning will occur, if it has not already occurred.

Will the opponents of reproductive human cloning, who maintain that it is harmful to a child to be brought it into the world in this way, say to such a child: "It is better that you were not born"?

Many of the arguments against reproductive cloning may also be applied to the engendering of children through sexual intercourse. For example, the claim that reproductive cloning is rooted in self-serving motives, in selfishness, in a quest for a type of immortality, could easily be applied to many parents of children brought into the world through sexual intercourse. Indeed, for centuries children have been brought into the world to supply labor, dowries, social status, or the genetic continuity of a family or a tradition. Certainly a human clone might have psychological obstacles to overcome, given the nature of how he or she came into the world. But why should we assume that such psychological problems would be any more severe than those faced by children in socioeconomically disadvantaged conditions, or those of children with chronic medical conditions? In this sense, *how* a child is brought into the world is not a decisive factor.

Ethicists worry about the "slippery slope" if cloning is not banned, but what about the "slippery slope" if it is? Will the criminalization of cloning lead to the criminalization of other forms of therapeutic and reproductive procedures?

The dangers of both reproductive and therapeutic cloning have been compared by opponents to the dangers of the atomic bomb and of creating genetically engineered pathogens that could wipe out human life on earth. Such comparisons appear overly alarmist and exaggerated. The production of a human being through cloning, or the production of stem cells through embryonic cloning, poses no overt danger to the continuation of the human species. When we take into account the possible destructive implications of many forms of scientific and techno-logical developments available today, the harm that might be done because of cloning pales in comparison. Yet the benefits from cloning may be immense. It aims at the creation of life, not its destruction; at the curing of a wide variety of diseases and medical conditions, not the creation of new maladies. Indeed, even a cloned Hitler could become a saint. Like any other human being, a cloned person would be more than the sum of his or her genes. Such an individual would be a unique, free moral agent in every way, a product of his or her unique experiences. Such an individual would be no more—and probably less—of a duplicate of his or her nucleic donor than would an identical twin. If risks deterred people from having children, how many people would there be today?

Good people make sensible arguments on both sides of the debate over cloning. Many of the most passionate advocates of cloning for biomedical research are individuals who themselves—or their children—might directly benefit from such research. Such individuals see restrictions on stem-cell research as a death sentence. For these individuals, any obstacle to aggressive re-search means unnecessary misery or death for many sufferers.

Perhaps we would not be so concerned about the possible abuses of cloning technologies if we could have a child in no other way, or if we loved someone whose life could be saved or immea-surably improved by the promised results of stem-cell research. One such person is Dr. Jack Kessler, a neurologist at Northwest-ern University in Chicago. After his fifteen-year-old daughter was paralyzed in a skiing accident, he put aside his research on nerve disorders and diabetes and devoted his attention to how future stem-cell therapies might repair spinal cord injuries. His dream is that one day biotechnology will allow his daughter to be able to walk down the street with his grandchildren.

A cost of foreclosing stem cell research is surrendering the opportunity of doing all that we can to improve the lives of the living—our own and those whom we love. Reflecting upon the opposition to human cloning for medical research, one scientist working in that field admitted that he feels as if his loved ones were trapped in a burning building—dying of diseases like diabetes and heart disease—and people are trying to tear from his hands a fire extinguisher: therapeutic cloning technology. In this regard it should be recalled that Dolly was cloned not primarily to show that mammalian cloning from adult stem cells could be done. The goal was to address the mandate to heal by using sheep to generate drugs for use in treating human diseases such as hemophilia and cystic fibrosis. Once cloned, sheep would be able to produce these drugs. In this and other cases, cloning technology cannot be cavalierly dismissed as an expression of human hubris aimed at showing that nature can be subdued. It is a means to an end that aims to improve the human condition.

Few theologians or ethicists have problems with the use of adult stem cells for biomedical research. What many of them find ethically problematic is the use of embryonic stem cells, because the embryo is destroyed in extracting these cells. For many theologians and religious bioethicists, the destruction of an embryo to obtain stem cells is morally wrong, whether the stem cells are taken from an embryo in an IVF clinic that might otherwise be discarded, or from an embryo created for the purpose of having its stem cells "harvested" and consequently destroyed.

In producing stem cells for research, an embryo is produced by means of cloning. At the early blastocyst stage of gestation, when the embryo is only a clump of cells, it is dismantled and the stem cells are removed, thereby destroying the cloned embryo. All extractions of stem cells from embryos involve their destruction. The use of embryos already created through IVF is considered by many to be less morally problematic than creating new embryos through cloning, because the new embryos are created only to be destroyed. In the case of existing IVF embryos, many would be discarded in any case. Yet many people still object. Some do so on theological grounds, maintaining that once conception occurs, the embryo is "ensouled" and has attained the status of human personhood. Others object on scientific grounds, arguing that unused embryos from infertility clinics are usually not as desirable for implantation and therefore not of the quality needed for research.

The issue of human cloning, either for reproduction or for biomedical research, presents a plethora of ethical problems. First and foremost is the status of the human embryo created through a combination of IVF and SCNT. At one end of the spectrum are those who consider an embryo little more than a cluster of cells with no more legal or moral status than any other group of human cells. At the other end of the spectrum are those who consider an embryo as having the same legal and moral status as children, adults, or other human beings. In this view they are the tiniest of human beings. But this approach sounds like—and may be influenced by—the medieval understanding of sperm, eggs, and embryos as a homunculus, an encapsulated miniature person. Some theologians add that since ensoulment begins with conception, an embryo has human moral and legal status from the moment of conception. This position, which considers an embryo at any stage of gestation to be a miniature human being, is for example one view endorsed by the President's Council on Bioethics. In its 2002 report, some members of the Council stated, "The embryo is in fact fully 'one of us': a human life in process, an equal member of the species *Homo sapiens* in the embryonic stage of his or her development." But Jewish law generally does not acknowledge the legal status of any entity invisible to the naked eye.

There is no doubt that human embryos are alive and that they are human, but so is each cell in a human body. Just as an embryo has the potential to develop into a human person, so too, through the use of cloning technology, does each cell of a human being. Research advances in cloning technology have in effect made each human cell "embryonic." Since each human cell may provide nucleic matter that could be used to clone a new human person, each human cell could be viewed as both "embryonic" and "ensouled." Would combing one's hair or scratching one's skin, either of which would involve the destruction of cells that could potentially become human beings, be considered a heinous deed by the opponents of cloning?

The argument that human life begins at conception is not technically applicable to cloning. In cloning there is no conception, which requires male and female genetic material derived from a sperm and an egg. In cloning one cannot accurately speak of conception or fertilization of the ovum since none occurs. Because there is no conception, the argument that since life and/or ensoulment begins at conception, destroying a cloned embryo is murder, becomes a moot point. In the sense that no conception takes place in reproductive cloning, a cloned embryo is like a golem: it is brought into being through artificial, asexual means.

In Jewish law, conception alone does not endow embryonic status. Indeed, scientists estimate that many eggs fertilized through sexual intercourse spontaneously abort and are flushed out of the woman's body in her menses. Certainly these fertilized eggs cannot be considered persons—though at one stage they may have had the potential to become persons.

For Jewish law, one characteristic of an embryo is implantation in the uterine wall for the purpose of gestation. Consequently a human blastocyst created in vitro—whether utilizing sperm and egg or utilizing cloning methods—would not be granted embryonic status until and unless it was implanted in a human uterus. Furthermore, Jewish law generally does not consider a fertilized egg as an embryo until at least forty days have passed since conception. Since embryonic stem cells are removed long before forty days of development, the entity from which they are taken is not yet considered embryonic. In this case, Jewish bioethicists accept a distinction often made by secular bioethicists between embryos and pre-embryos.

Pre-embryos originally created through IVF for reproductive purposes, that will be discarded in any case, not only could but should be used for stem-cell research, if the appropriate legal releases for such use are obtained. Certainly it is better to use such already existing pre-embryos for potentially lifesaving and therapeutic reasons than simply to discard them or freeze them to await eventual destruction. It is estimated that at least 400,000 such embryos are currently being stored by assisted-reproduction clinics.

Jewish law would find the use of adult stem cells or stem cells derived from pre-embryos created through IVF (and about to be discarded) to be preferable to the use of stem cells from pre-embryos created by cloning for the express purpose of extracting stem cells. But this preference would in no way preclude the use of embryonic stem cells for biomedical research. Throughout the history of Judaism, the mandate to heal and to preserve existing human life at almost any cost has been highly valued. The potential benefits of stem cells, including embryonic stem cells, would certainly fall under that mandate.

What are the ethical implications of granting an embryo one status or another where cloning is concerned? The claim that a human pre-embryo or a human embryo has full human status means that anyone taking part in its destruction would be guilty of murder. But if a human pre-embryo, or even a human embryo, does not enjoy the status of human personhood, what status does it have according to Jewish law? The dominant view in Jewish law, from the Bible onward, has been that it is property. The Bible (Exodus 21:22) requires a person who accidentally injures a pregnant woman, thereby causing her to abort, to pay tort and property damages. This indicates that the embryo is considered to be property and not a human being with the moral status and legal protections that would imply. In American law a similar view has been taken. In its "Ethical Statement on In Vitro Fertilization," the American Fertility Society specifically refers to sperm, eggs, and embryos as the property of the individuals who provide them.

Yet a second option presents itself, and that is to consider an embryo as having the status of a golem. In the passage from Cynthia Ozick's novel with which this chapter began, the golem in the story describes herself as feeling like an embryo when she came to life. Considering a human embryo as a golem would give it a unique status, with clear implications for human reproductive cloning, cloning for biomedical research, and abortion.

In his definitive study of the golem in classical Jewish literature, Moshe Idel established that *golem* was often understood in embryonic terms. Whereas in modern Hebrew *golem* is the word for "cocoon," in medieval Hebrew it was often used to denote an embryo, one that has developed far beyond the blastocyst stage. Golem can also denote an embryo with clearly defined limbs gestating in a human uterus. In a tenth-century Hebrew poem, Amitai ben Shefatiya writes of "The texture of the limbs of the *golem* [the embryo] in the womb, when it was formed."

As noted earlier, rabbinic legends describe the creation of Adam as, first, a golem—with human limbs and organs. Only later, when Adam is granted a soul, does Adam-Golem become Adam-Human. Hence, as Idel points out, in early rabbinic literature the "Golem is understood as the embryonic state." The golem represents the physical blueprint for a complete human being, though it is not yet itself a human being. And we have seen that some classical Jewish scholars maintain that, like an embryo, a golem might become a human being and acquire the legal status of a human. Like a human embryo, a golem is potentially a human being; but while it retains its golemic status it cannot enjoy the legal or moral protections of personhood. In this view, a golem, like an embryo, has neither the status of a mere clump of cells nor the status of a person. Rather, it has its own discrete legal status. Similarly, in a case in Tennessee (*Davis v. Davis*) that concerned the disposition in a divorce settlement of frozen embryos initially created through IVF, the court considered the embryos to have a unique status of their own, neither a person nor property.

Recalling Rabbi Zevi Ashkenazi's legal opinion that a golem may be destroyed because it is not a human being, we can see that the status of personhood is too precious a commodity to be bestowed precipitously. It carries with it unique legal and moral protections. Thus golems and embryos should not be granted the status of persons, because they are not. They are what they are, not more and not less. According to a prominent bioethicist, embryos are neither "people nor products."

Because embryos, like golems, have a legal status of their own that distinguishes them from persons, they may be created as well as destroyed. There needs only be a viable purpose for doing so, particularly a purpose aimed at the improvement of the life and health of existing human beings—as, for example, in the case of using embryonic stem cells for therapeutic cloning. But if a golem or a cloned human embryo attains the fundamental characteristics and status of a person (as in the birth of the originally cloned embryo), he or she becomes legally and morally indistinguishable from any other human being.

According to Zevi Ashkenazi and his son Jacob Emden, a golem is not a human being, nor does it have the potential to become one. Other rabbis disagree about its potential. Still, potentiality is not actuality. Neither a golem nor an embryo enjoys the legal status of a person. Neither should be treated as a person because it has not developed all the necessary features of personhood, and may never do so. Just as an acorn is not an oak tree, neither a human embryo nor a golem is a person. Potential alone does not confer personhood. Each sperm, each egg, each embryo, and perhaps each golem has the potential for personhood, but most never realize that potential.

Opponents of cloning and other reproductive biotechnologies maintain that such methods are unnatural because they deviate from the "natural" method of human reproduction—they separate procreation from the sexual act. Beginning with artificial insemination and moving to in vitro fertilization and the possibilities of human reproductive cloning, human reproduction has become increasingly detached from sexual intercourse.

Although Jewish religious law has permitted the use of most methods of reproductive biotechnology, other religious traditions consider these methods to be unnatural and immoral. *The Catechism of the Catholic Church* (paras. 2376–2377), for example, states, "Techniques that entail the disassociation of husband and wife, by the intrusion of a person other than the [married] couple, are gravely immoral. . . . Techniques involving only the married couple are perhaps less reprehensible, yet remain morally unacceptable. They disassociate the sexual act from the procreative act . . . and establish the domination of technology over the origin and destiny of the human person. . . . Only respect for the link between the meanings of the conjugal act and respect for the unity of the human being make possible procreation in conformity with the dignity of the person." Similarly, Leon Kass opposes various methods of reproductive biotechnology on the grounds that it is an immoral affront to human dignity and identity. With specific reference to various forms of human asexual reproduction, like reproductive cloning, Kass declares, "To say 'yes' to asexual reproduction and baby manufacturing is to say 'no' to all natural and human relations, is to say 'no' also to the deepest meaning of coupling, namely, human erotic longing." For Kass, such methods are unnatural to humans and reflect the sexuality of lower forms of life, like bacteria or fungi. Yet, according to Scripture, the ancestors of the human race, Adam and Eve, each come into being through asexual reproduction. The New Testament considers Jesus to have been conceived through asexual reproduction. As we have seen, the golem, whether created by God or by humans, is also a product of asexual reproduction.

From a Jewish perspective, bringing creatures into the world through asexual reproduction is an act of *imitatio Dei*. What is important is not so much the method of reproduction but the purpose for which it is employed. If that purpose is the creation of a pre-embryo that has the potential to help save life and conquer disease, or if the purpose is to provide a child for someone who otherwise might never become a parent, it is both permissible and desirable.

According to the Bible, the first person brought into being through the use of adult stem cells was Eve, the mother of humankind, who was created from one of Adam's limbs. Like Adam, she came into the world through an act of asexual reproduction. Adam, as we have seen, was first created as a golem who later became a human being. Similarly, the golems we can create today through various methods of reproductive biotechnology often have the potential of becoming persons. But while in the golemic state, they retain the status of golems.

According to Zevi Ashkenazi, a golem is not granted human status because it has never been implanted in and delivered from a human womb. Like a golem, a pre-embryo cloned for biomedical research is a product of asexual reproduction. Yet Rabbi Gershon Hanokh Leiner of Radzyn rejected Ashkenazi's view that a golem could not be granted human status simply because it is not "of woman born." More important, says Leiner, is the golem's subsequent status and identity. The eventual status of the individual, rather than the process by which that individual was conceived or gestated, is the crucial factor.

Once born, a cloned human being would be human in every sense, with all the moral and legal implications that entails. Genetically the clone would be a close facsimile of the nucleic donor, but not an exact duplicate. Indeed, identical twins would be closer duplicates of one another than a clone and its nucleic donor, because in the case of the clone, the mitochondrial cells (the "white" of the egg in which the nucleus is placed) would have genetic materials of their own. And the experiences, social conditions, and education of the cloned person would be different from those of the nucleic donor.

In the summer of 1973, George Hudock, then a professor of zoology at Indiana University, published an article in the *Indiana Law Journal* in which he stated his opposition to human cloning. Nonetheless Hudock wrote that "except for gametes, a human being is a clone because all of its cells are derived from a single fertilized ovum. Moreover, identical twins, those derived from one fertilized ovum, are a clone." Put another way, we are all clones. Cloning regularly occurs in nature. Cloning by human beings, therefore, imitates nature and is not unnatural.

Similarly, according to Judah Loew, the creation of a golem is not done by artificial means because the method of creating a golem is not outside the "natural order." Genetic engineering too might be considered within the natural order. In 1982 the President's Commission for the Study of Ethical Problems in Medicine and Biomedical and Behavioral Research declared: "The basic processes underlying genetic engineering are natural and not revolutionary." While a golem is created from dust rather than from genetic materials, both golems and human beings ultimately return to dust. "Dust you are and to dust you return," God reminds Adam. Golems and humans share a common origin and a shared final destination. Neither is outside the natural order.

None of the medieval commentaries to the Talmudic story of the creation of a golem found it to be theologically or ethically problematic. One fourteenth-century commentator, Rabbi Menahem ha-Meiri, noted with approval that "There will come a time when science will know how to create human beings without the intimate act. This has been explained in the books of science and is not an impossibility. . . . Even the knowledge to create new beautiful creatures that are engendered through asexual reproduction, as is noted in the books about natural science, cannot be proscribed."

Many writers on biotechnology in general and on human cloning in particular have compared these activities to those of the protagonist of Mary Shelley's novel *Frankenstein*. Those who identify human cloning as intrinsically immoral, unnatural, reprehensible, inevitably harmful, destructive, and as "playing God" seem to adapt a perspective shared by and perhaps inspired by *Frankenstein*. They view both the methods and implications of the artificial creation of human life as a threat to human identity, a perversion of human power and creativity, a violation of human dignity, an expression of human hubris, an inevitable catastrophe waiting to happen. But the legend of the golem and the Jewish religious tradition from which it comes offer a radically different view. They affirm that the artificial creation of human life is sanctioned within certain spiritual, ethical, and legal boundaries. When purposeful and beneficial, such activities are seen as an imitation of God, not as "playing God." Rather than a usurpation of divine prerogatives, they are considered the fulfillment of a divine mandate to continue God's work of creation, to serve as "God's partners in the work of creation." As such, these activities are perceived as extensions of nature rather than the unnatural practices of the "unhallowed arts" pursued by Victor Frankenstein. Although the golem legend recognizes the risks of creating artificial life, both to the creator and to the creature, it advises prudent and responsible caution and does not presume inevitable catastrophe.

The idea of "playing God" is not unknown in Jewish tradition. Pretending to be God, or incorrectly identifying something or someone other than God with God, is idolatry, the Bible's most heinous sin. Judaism has a long history of distinguishing between human beings and God, and of warning human beings against trying to impersonate God. But Judaism also has a long history of encouraging human beings to imitate God by sharing God's attributes of justice and mercy, love and compassion. Human beings articulate their having been created in the "image of God" by imitating the most uniquely divine attribute of all: creativity. As we have seen, for the Talmudic rabbis and the medieval Jewish mystics this includes the creation of life. For Judah Loew of Prague, human creativity was a natural way of transcending and extending nature beyond its original limits.

Jewish tradition clearly distinguishes between two types of creativity: the creation of something from something, and the creation of something from nothing. God and human beings share in the ability to create something from something, to create something new from something that has already been created by God. Such human creativity is desirable, as it imitates God. But only God can create something from nothing. Humans cannot. In this view, "playing God" applies only to situations in which the human being claims to create something from nothing, in which the human creature fails to acknowledge the divine source of his or her creative abilities, in which the human being mistakenly equates his or her own power with divine omnipotence, in which human creative accomplishments engender arrogance and pride rather than gratitude and humility. Human creativity must rest content with the extraordinary gift of being able to create something from something: bread from wheat, linen from flax, sculptures from marble, new life from existing life, golems from dust.

Reproductive biotechnology does not "create" life. It does not create something from nothing. Rather, it creates something from something. By using nature's raw materials, it aims to engender life and improve the quality of existing life. Meanwhile the creation of life itself, the creation of life "from scratch," remains a prerogative vouchsafed only to God. Only when human beings seek to usurp that prerogative does Jewish tradition consider them as playing God. As an early rabbinic saying puts it, "If all the creatures of the world gathered together to make a single gnat from scratch and to infuse a soul into it, they would not succeed." In this regard, consider the following story:

Once a group of scientists wished to demonstrate that since they had acquired godlike abilities through scientific and technological advances, God was now superfluous. To demonstrate this, they challenged God to a contest. Each would create a human being from a lump of clay.

Two lumps of clay were set out in a laboratory, and a team of scientists began the process of creating a human being from it. Meanwhile the assembled audience waited for something divinely initiated to happen to the other lump of clay.

The scientists worked on, trying to transform their lump of clay into a human being. Suddenly, however, a bright light filled the room, and their lump of clay disappeared. A voice was heard to say: "Before creating a human being from clay, each contestant must first create his or her own clay."

Chapter Eight

Mechanical Golems: Toward a Postbiological Human Future?

Rabbi Solomon ibn Gabirol was the greatest Hebrew poet of the Middle Ages and a prominent philosopher. His philosophical works, written in Arabic, influenced subsequent Jewish mysticism, Christian scholasticism, and modern astrology. His Hebrew poetry is still read today, and his liturgical poems are currently recited in synagogues around the world. Many legends about ibn Gabirol have been preserved, and according to one of them, ibn Gabirol created a golem—different from any golem that had been created before.

Ibn Gabirol lived in eleventh-century Spain; he died while in his thirties. Orphaned at a young age, he was small of stature, physically weak, and considered himself ugly. He suffered from a severe skin disease, probably elephantiasis, which was thought in his day to be both contagious and fatal. So he was often alone, sometimes in involuntary quarantine.

Although he composed many passionately romantic poems, ibn Gabirol's physical condition probably precluded female attention and companionship. It should not be surprising, therefore, that according to one legend he created a female golem to be his companion and housekeeper. As the story goes, when word of this golem's existence reached the civil authorities, they presumed that ibn Gabirol had created the golem for sexual purposes. So they ordered him to destroy it. Ibn Gabirol complied, and the golem returned to the wood and hinges from which it had been created.

The legend of ibn Gabirol's golem establishes two precedents in classical Jewish literature. His was both the first female as well as the first mechanical golem. Unlike golems before her, ibn Gabirol's golem was not created from clay. Made from wood and hinges, ibn Gabirol's golem was a rudimentary machine.

Today machines—mechanical golems—populate our homes, our workplaces, our daily lives. Like golems, machines are artifacts that we create to help, serve, and defend us. Like some golems, however, some of these machines have the capacity not only to protect us but also to harm and even to destroy us. Increasingly in our daily lives we coexist with machines.

Machines have become our constant companions, indispensable extensions of ourselves. We have become progressively dependent upon them. From the alarm clock that awakens us in the morning, to the coffeemaker that prepares our first daily cup of coffee, to the television we watch before going to sleep, machines serve us throughout each day. From the first ultrasounds of our fetal images months before we are born, to the brain and heart monitors that record our last heartbeat and certify our death, machines are constantly with us. We could hardly imagine our daily lives without them. Although a person may live alone, he or she would be hard pressed to live without machines.

Among the myriad of machines, computers have become pervasive in our lives. We have grown increasingly dependent upon them. Not so long ago, if all the computers in the world had stopped functioning, the disruption to our daily lives would have been minimal. Today—and certainly in the future—were all the computers in the world to crash, the impact on our lives would be nothing less than catastrophic. Human activity would grind to a silent, horrifying halt. Electrical power distribution would fail. The microprocessors found in most motor vehicles would stop functioning, and our cars would not run; neither would most trucks, buses, airplanes, or trains. Telecommunications would be silenced: telephones, radios, fax machines, pagers, palm pilots, and personal computers would shut down. Government agencies, banks, and police and fire departments would be able to provide only minimal services. Economic chaos would ensue. No one would be able to use a credit card or an ATM machine; financial transactions would largely stop; financial markets would be compelled to shut down. And if all the data on all the computers were suddenly to vanish, or if even parts of memory banks were to be wiped away, it would be a cataclysm with long-term effects.

In 1964, Norbert Wiener, in his book, *God and Golem, Inc.*, identified "the relation between man and the machine" as "one of the great future problems we must face." For Wiener, many dimensions of that problem had already been anticipated by the golem legend. As Wiener reminds us, "The machine . . . is the modern counterpart of the Golem of the Rabbi of Prague."

In 1965, the year after Wiener's book was published, the savant of Jewish mysticism and author of scholarly studies on the golem, Gershom Scholem of The Hebrew University of Jerusalem, spoke at the installation of a new computer at Israel's Weizmann Institute. In his remarks Scholem described the computer as a new form of golem. After drawing numerous parallels between the golem and the computer, Scholem observed that the "new golem"—the computer—seemed to be superior to the old golem: it could learn from its mistakes and could improve itself.

Like computers and other machines, robots and androids have been described by artists, scholars, and scientists as the golems of our times. When the Cepak brothers coined the word *robot* in their 1921 play *R.U.R.*, it is likely they were influenced by the legend of the golem of Prague, which was well known in their native Bohemia. As robotics, artificial intelligence, and computer science develop, these "new golems" will become more like us. At the same time we may become more like them.

In his book *Flesh and Machines: How Robots Will Change Us*, Rodney Brooks, a major authority in robotics and director of MIT's Artificial Intelligence Laboratory, examines the relationship between humans and robots. Brooks writes, "Mankind's centuries-long quest to build artificial creatures is bearing fruit. Machines are now becoming autonomous in areas that bypassed them in the industrial revolution. . . . We are starting to see intelligent robots. . . . But these robots are not just robots. *They are artificial creatures* [emphasis added]. Our relationships with these machines will be different from our relationships with all previous machines. The coming robotics revolution will change the fundamental nature of our society. . . . Our machines will become much more like us, and we will become much more like our machines. . . . While we have come to *rely* on our machines in the last fifty years, we are about to *become* our machines during the first part of this millennium. We . . . will morph ourselves into machines. . . . The distinction between us and robots is going to disappear." Or, as Bill McKibben puts it in his book *Enough: Staying Human in an Engineered Age,* the problem now confronting us is: "Will humans be turned into robots before robots can be turned into humans?"

According to cutting-edge scientists in robotics, computer science, and artificial intelligence, like Rodney Brooks, Hans Moravec, and Ray Kurzweil, we can anticipate a future in which humans and machines will merge into a single species, where they will become indistinguishable from one another. In that future, machines will acquire human status and human beings will become machines. Put another way, not only will golems become human, but humans will become golems. Although today's robots have very far to go to be anything like Data of "Star Trek" or David of the film *AI,* scientists as well as theologians are already contemplating a fast-approaching future in which robots and humans will become alike in most important ways. Even now, religiously committed computer scientists like Edmund Furse are suggesting that robots of the future could be baptized and ordained to the priesthood. Others are speculating about whether humans and androids might marry one another.

In the nineteenth century the British novelist Samuel Butler considered the possibility that machines might be the next step in human evolution, that human beings could become a race of machines. For Butler this possibility was so horrifying that he suggested destroying all machines to preclude such a possibility from ever happening. Unlike Butler, a number of contemporary scientists look forward to a postbiological human future, to a time when life on earth—including human life—will be silicon based, when we will be able to download our consciousness and knowledge onto computers. In this postbiological human future, our organic bodies would become obsolete. We would exist either as machines or as virtual entities in cyberspace.

The technology that drives these kinds of predictions and aspirations is new. But even in the early seventeenth century, the claim that the universe and human beings are machines had begun to take root in Western culture. In ancient and medieval times the universe was described as a huge living organism. But by the seventeenth century the universe was understood to be an enormous machine that ran with the precision of a well-crafted clock. As the astronomer Johannes Kepler wrote at the end of the sixteenth century, "My aim is to show that the Celestial machine is to be likened not to a divine organism but rather to clockwork." During the Middle Ages, human beings were considered worlds in miniature—microcosms. But once the universe was conceived to be a machine, human beings began to be described as miniature machines.

The problem of whether humans are merely machines was addressed by the seventeenth-century French philosopher René Descartes, considered by many historians of philosophy to have been the first significant modern philosopher. Descartes dealt with this problem by trying to demonstrate that only part of the human being—the human body—is a machine, albeit a flawed and unreliable one. For Descartes, however, the essence of the human being is not the body but the soul or mind. It is the mind that makes us human and uniquely different from the universe and from other living creatures. And, though the body is a machine, says Descartes, it is directed by the mind. Animals, on the other hand, have neither minds nor souls. They are only bodies, just machines. As such they have neither intelligence, nor feelings, nor personalities. Descartes's claim that animals are merely unfeeling machines was used to justify horrific experiments on animals, including live, unanaesthetized vivisections.

For Descartes, animals are "beast-machines." Humans, however, are machines with intelligent minds and immortal souls. Like Rabbi Jacob Emden who considered golems as a form of animal and therefore without legal protection, Descartes considered "beast-machines" as dispensable. Emden's justification of Rabbi Zeira's killing of Rava's golem, on the grounds that the golem's inability to speak demonstrated that it was not human, parallels Descartes's observation that since dogs cannot speak French, they (like golems) have neither intellects nor souls and therefore can be destroyed.

Descartes died in 1650. Almost a century after his death, in 1748, another French philosopher, Julien Offray de la Mettrie, published *Man, A Machine*. Unlike Descartes, who considered the workings of the mind to be a function of the soul, La Mettrie considered human intellectual activities to be physically rather than spiritually based. He explained all animal behavior, including human behavior, in crassly mechanistic terms. Since for La Mettrie there was no human soul, human beings—like other animals—were only bodies, or machines. Nonetheless La Mettrie felt obliged to insist that human beings somehow differed from other animals—that is, from other living machines. Unlike other machines, said La Mettrie, human machines "wind their own springs."

The philosophical approach known as "physicalism," and represented by philosophers such as La Mettrie, was perpetuated and amplified during the nineteenth and twentieth centuries. The prominent nineteenth-century American philosopher and politician Robert Ingersoll described the human being as "a machine into which we put what we call food and produce what we call thought." In other words, humans are thinking machines. In the early twentieth century a popular definition of the human being was "an ingenious array of portable plumbing." Isak Dinesen, the author of *Out of Africa*, once wrote, "What is man when you come to think about him, but a minutely set, ingenious machine for turning with infinite artfulness, the red wine of Shiraz into urine?"

Once internalized, the conception of the human being as a machine influenced self-conceptions that came to be expressed in colloquial speech. People began to speak of themselves in mechanical terms. For example, we readily describe ourselves as being "turned on" and "turned off." We "tune in" and "tune out." We provide "input," "output," and "throughput." We go on vacation to "recharge our batteries." We "gear ourselves up" and "get our motors running." Sometimes we are too "wound up" and must "wind down." A dying patient is often described by physicians as having a "systems shutdown."

The view of the human being as a soulless machine approximates the description of ibn Gabirol's golem. But it also reflects a dominant conceptual model embraced by much of twentieth-century American medicine. The psychiatrist George Engel has called this the "biomedical model." A feature of this approach is to consider a sick human being primarily as a machine in need of repair.

While the seventeenth-century paradigm of the machine was the well-crafted clock, by the end of the twentieth century it became the computer. Today scientists like Stephan Wolfram regard the universe as an enormous computer that runs certain programs. If we could understand these programs, we would understand how the universe works.

While in the nineteenth century life was viewed as an aggregate of separate and interchangeable parts assembled into a working whole, by the end of the twentieth century life was viewed as a code—the genome, consisting of many bits of information capable of being programmed in specific ways.

Once life was understood to be an information system embodied in a genomic code, life became a code to crack. The tool to be employed in cracking this code became the computer. It is not coincidental, therefore, that one of the greatest "code crackers" of all times was also one of the founders of computer science. His name is Alan Turing.

During World War II, Turing worked for British Intelligence. He was largely responsible for the breaking of German military codes that immeasurably helped the Allies during the war. If the German "Enigma" code had not been broken by Turing and his assistants, England might well have lost the Battle of Britain and perhaps the war as well. To break the complex German codes, Turing and his colleagues constructed the first operating computer, which he later replaced with more sophisticated machines of his own design. After the war, in 1950, Turing published his now classic study *Computing Machinery and Intelligence*, in which he set out the agenda for computer research for the next five decades. In this study Turing begins by addressing the question, Can machines think?

To answer the question, Turing established the now famous "Turing Test." When a person communicates with another party over a keyboard and cannot discern whether he is communicating with another person or a machine, it is appropriate to claim that computers can think, when the other party is in fact a computer. Scientists today debate whether contemporary computers are capable of passing the Turing Test.

Like many scientists today, Turing considered human beings to be machines, "human computers." But while many contemporary computer scientists claim that computers can actually think, and reject the notion that computers merely mimic "human computers," Turing claimed that thinking machines would be those that "do well in the imitation game." Looking toward the future from 1950, Turing wrote: "The original question, 'Can machines think?' I believe to be too meaningless to deserve discussion. Nevertheless, I believe that at the end of the [twentieth] century the use of words and general educated opinion will have altered so much that one will be able to speak of machines thinking without expecting to be contradicted." Turing predicted that one day computers would compose poetry and beat chess masters at their own game.

When Norbert Wiener first formulated the field of cybernetics in the 1940s, he claimed that the operating principles of information processing could be extended from engineering to biology, that animals (including human beings) as well as machines could readily be described as "information systems." By the 1960s biologists were readily describing life in cybernetic terms. By the 1980s molecular biologists were speaking about living cells as machines. In his 1985 book *The Origins of Life,* the physicist Freeman Dyson wrote, "Hardware processes information; software embodies information. These two components have their exact analogues in the living cell; protein is hardware and nucleic acid is software." Similarly, by the late 1980s standard biology textbooks had been rewritten to reflect the pervasive influence of the information sciences on the biological sciences. In one such popular textbook, *The Molecular Biology of the Cell,* we read: "For cells as computers, memory makes complex programs possible; and many cells together, each one stepping through its complex developmental control program, generates a complex adult body. . . . Thus the cells of the embryo can be likened to an array of computers operating in parallel and exchanging information with one another."

Much of biology now seems as if it has become a branch of engineering or computer science. As one computer scientist has put it, "Biology, particularly at the molecular level, can be viewed for many purposes as an information science."

Further integration of the information and biological sciences will come in the form of a "molecular computer," made of DNA rather than of silicon. In 1996 scientists produced the first DNA chip. In his 2002 book *Machine Nature: The Coming Age of Bio-Inspired Computing,* Moshe Sipper of Israel's Ben Gurion University writes about the use of "DNA molecules as building blocks for a novel type of computer." In "DNA computing," rather than utilizing the binary 0-1 number system, the abbreviation of GATC, representing the nucleotides of DNA, will become the basis of computer language. Such DNA-based computers will be living entities.

Reflecting on recent developments in biology and computer science, an MIT historian of science has observed that "The body of modern biology, like the DNA molecule—and also like the modern corporation or political body—has become just another part of an informational network, now machine, now message, always ready for exchange, each for the other."

It should not be surprising that more recent developments in biology, and especially in genomics, have witnessed the convergence not only of biological research and computer science but also of biological discourse and computer jargon. Biologists used to describe the actions of organisms as "behavior"; many now use terms drawn from engineering and computer jargon, such as "performance" and "efficiency." These words also have become standard in corporate discussions of economic conditions and employee behavior. In other words, corporate bodies as well as human bodies are now understood to be little more than types of information systems. We no longer describe ourselves as generic machines; now we are more specific. Like Turing, we describe ourselves as a type of computer, a "human computer."

The project to publish the "code of life," the human genome, required the development of a generation of supercomputers. Without these machines the human genome would not have become available so quickly, nor would many developments in bioengineering or "artificial life" have been possible. So, in a real sense, science has become computer science. As the biological sciences and computer science converge, as biology and engineering merge, some scientists have begun to look forward to the time when, in Ray Kurzweil's words, "there is no longer any clear distinction between humans and computers." Kurzweil expects this merger of humans and computers to occur by the end of the twenty-first century.

Scientists readily identify the activities and functions of human organisms with those of machines. Rodney Brooks can dedicate his book *Flesh and Machines* to his "darling wife," then write: "The [human] body, this mass of biomolecules, is a machine that acts according to a set of specific rules. . . . The body is a machine. . . . We are machines, as are our spouses, our children, and our dogs. . . . Every person I meet is also a machine— a big bag of skin full of biomolecules interacting according to describable and knowable rules. When I look at my children, I can, when I force myself, understand them in this way. They are machines interacting with the world."

To bring about the eventual merger of humans and machines as the next step in human evolution, two things have to happen, according to scientists like Brooks and Kurzweil. Humans must become more mechanical, and machines must acquire characteristics thought to be particularly human. From this perspective, a crucial step in the transition of humans to machines will occur when humans become part human/part machine, or cyborgs. *Cyborg* is a combination of two words: *cybernetic* and *organism*. In his influential 1970 work *Future Shock*, the sociologist and futurist Alvin Toffler predicted that "The 'Cyborg'—which is the name . . . for animal-machine combinations—seems to be the man of the future." For many, that future has arrived.

Scientists like Kurzweil believe that accelerating developments in bionics, computer science, biotechnology, and robotics will speed this transition from humans to human-machines. We will increasingly integrate mechanical devices into our bodies until we are completely machines. Already, these scientists point out, people have pacemakers embedded in their bodies to regulate cardiac function, titanium joints to help them walk, cochlear implants to enable them to hear, insulin pumps to treat diabetes, and soon silicon retinal implants to improve sight. Artificial organs and limbs, currently in the experimental and developmental stage, will inevitably be employed. Tiny nanorobots will dwell inside our bodies diagnosing illness and making repairs. Electrodes and computer chips will be implanted to stimulate damaged nerves, improve memory, and treat various diseases and conditions. Computers of various sizes will not only be carried about by us but will be embedded in our clothing as well as throughout our bodies. Eventually the human brain will become integrated with or replaced by manufactured neurocircuits, connected to data bases and computers around the world.

From this perspective, we will eventually and inevitably become products of our own technology. Unlike previous steps in human evolution, these advances will be chosen by human beings and will no longer be the product of chance. We will, in effect, create the next edition of the human species. As Kurzweil puts it, "The next inevitable step is a merger of the technology-inventing species with the computational technology it initiated the creation of. At this stage in the evolution of intelligence on a planet, the computers are themselves based at least in part on the designs of the brains (that is, computational organs) of the species that originally created them and in turn the computers become embedded in and integrated into the species' bodies and brains. . . . The Law of Accelerating Returns predicts a computer merger of the species with the technology it created."

The transition of humans into machines is being paralleled by developments aimed at making machines indistinguishable from human beings. As these developments unfold, the gap between humans and machines will close. The acquisition by machines of characteristics previously thought to be uniquely human is forcing us to consider how we are distinct from the machines we have created, to reconsider the nature of human nature.

Here are characteristics considered to be uniquely human by a wide variety of philosophers, theologians, and scientists who have confronted this issue: life, intelligence, thought, consciousness, creativity, reproduction, emotion, embodiment, speech, a sense of humor, individual identity, moral autonomy, a soul, free will, and the ability to blush, to learn, to lie, to move, to decide, and to discover new things.

In science fiction literature and film, we encounter robots, androids (robots in human form), and computers that have many if not most of these characteristics. Data in "Star Trek" possesses many of these features. Although he has no emotions and may not have a soul, Data nonetheless exhibits many particularly "human" characteristics in a manner that often surpasses human capabilities. Robots, such as *Star Wars*'s C3PO, not only express emotion but also neuroses. David in Steven Spielberg's film *AI* seems human in most ways.

Robots, androids, and computers exhibit both desirable and undesirable human traits and activities. Hal in Stanley Kubrick's *2001: A Space Odyssey* expresses jealousy and becomes a murdering psychopath. In the 1977 film *Demon Seed*, a computer "rapes" Susan Harris (played by Julie Christie) so as to have offspring (in this case, a cyborg).

Stephen Hawking has said that "Today's science fiction is often tomorrow's scientific fact." In the 1950s, for example, science fiction portrayed computers that could defeat human beings at chess. Since the ability to play chess was thought at that time to be a uniquely human ability, scientists relegated such scenarios to the realm of fantasy. But by the late 1990s, even the world's reigning grand master in chess, Garry Kasparov, was beaten by a computer. Similarly, not so long ago, areas such as linguistic syntax were considered a uniquely human domain. Today our personal computers check our spelling and our grammar, and even make "recommendations" about how to employ syntax. If they provide us with information in a language unknown to us, they often can translate it for us. Computers provide us with information as well as "advice." They have become our "teachers"—sometimes even our psychotherapists.

Using a simple syntactical program, in 1963 Joseph Weizenbaum at MIT created a program called ELIZA. Its purpose was to demonstrate that computers should not be confused with people, that computers cannot think, that computers would inevitably fail the Turing Test. To his horror and surprise, Weizenbaum found that some people would spend hours revealing their most intimate problems and experiences to the computer, speaking to it as if it were a human psychotherapist.

Computers now compose music and poetry, as Turing thought they would. They predict and plan for a wide variety of future phenomena and events, including everything from weather to financial trends, and strategies for military conflicts yet to occur. Computer scientists like Rodney Brooks anticipate a time in the not too distant future when androids, robots, and computers will express emotions, tell jokes, have consciousness, self-reproduce, and become autonomous automatons. When these things occur, says Brooks, distinctions between humans and machines will disappear.

Besides anticipating future scientific and technological developments, science fiction also reflects our current hopes and fears about the unfolding future. Futuristic scenarios that articulate our hopes also express our apprehensions. Benevolent, even servile androids and computers often yearn to become human like "Star Trek's" Data and *AI*'s David; yet by giving these artificial life-forms human aspirations, we suppress our fears that such humanly created artifacts might eventually surpass, replace, and control us. We try to reassure ourselves that human beings will remain superior to and in control of the artifacts they create. In fact most science fiction offers us a bleak, almost apocalyptic vision of the future, when machines will control or seek to control us, when we shall have surrendered our human nature and become like them. As in many recent versions of the golem legend, the story line in many of these science fiction stories is identical: the entities we have created to help and defend us, eventually threaten to harm, control, or destroy us. Surpassing us in power and often in intelligence, we are no match for the creatures we have introduced into our world. In films such as *Colossus: The Forbin Project*, *Terminator 3: The Rise of the Machines*, and *The Matrix*, apprehensions about a mechanistic future are graphically portrayed.

Probably the first literary work to examine a future in which machines surpass and replace humans is *R.U.R.* The letters form the abbreviation for "Rossum's Universal Robots," a conglomerate that manufactures androids. *R.U.R.* describes a time when human beings have created androids as servants. The goal of their creation is to guarantee economic prosperity and allow humans the opportunity to pursue the perfection of their own species. Rather than improving themselves physically and spiritually, however, humans become vacuous, lazy, and complacent. In the meantime, the androids adapt many human virtues and eventually exceed human beings in intelligence, physical endurance, and a sense of purpose. And the androids soon conclude that human beings have no purpose, that they have become "superfluous, unnecessary." Consequently the androids rebel and destroy their creators. These "artificial people," as Cepak calls them, both surpass and replace the human species.

Asked by one of the last remaining human beings, "Why did you murder us?" the leader of the android rebellion in *R.U.R.* responds that in aspiring to become like humans, androids have acquired not only human virtues but also human vices. In the words of Radius, the android leader: "Slaughter and domination are necessary if you want to be like men. Read history, read the human books. You must domineer and murder if you want to be like men." By adapting human characteristics, the androids have acquired emotions and souls. Radius continues, "But terror and pain turned us into souls. . . . We feel what we did not used to feel. . . . Teach us to have children so that we may love them." At the play's end, a male and a female android fall in love and emerge as the Adam and Eve of a new humanly created species that will rule the earth. Human beings are annihilated by the very creatures they had created to help, serve, and defend them.

According to some of our leading scientists and anthropologists, the transition from humans to machines already has begun with the introduction of bionics. Our transformation into cyborgs is viewed by contemporary scientists as a critical step toward our eventual future as postbiological human machines. Yet we need not consider ourselves cyborgs simply because medical technology has introduced mechanical devices that can improve our health and quality of life. Just because grandma has a pacemaker, an artificial hip, implanted teeth, an insulin pump, and a cochlear implant, she is not a cyborg or a "robosapien."

Nor are we obliged to accept the claims that machines possess essential human characteristics, such as human intelligence. Even the great pioneer of computer science, Alan Turing, admitted that machines do not possess human traits but only mimic them. Twentieth-century philosophers such as Michael Polanyi and William Barrett have argued that machines do not have the human characteristics that some ascribe to them. Rather, these characteristics are really only reflections of the virtues and vices of their human creators. In this view, such creatures tell us more about ourselves than about them, more about our nature than theirs. Reflecting on Turing's tortured and unhappy life, which ended with his suicide in 1954 at the age of forty-one, Barrett writes, "For a man whose mind had been continuously engaged with the question of how the machine might guide and regulate life, he seems to have been sadly incapable of managing his own."

At the root of the long-standing claim that humans are machines is an approach that defines things by their functions rather than by their nature. From this perspective it follows that if the universe, the organic molecule, and the human being function like a machine, they must be machines. Thus information systems like computers have become the functional model for the universe and for all life, including human life. But just as the universe is not a clock, as seventeenth-century scientists claimed, neither is it a computer.

If we begin with the assumption that the universe is a machine, whether a clock or a computer, it is all but inevitable that we will understand ourselves as machines. Once machines become the paradigms of existence, once computer-based concepts and jargon serve as the basis for explaining and describing all natural phenomena, it should not be surprising that we begin to view all things—including ourselves—as computers. But just as it is a mistake to explain the universe in anthropocentric terms, so is it a mistake to impose upon all natural phenomena categories of thought and speech derived from computer science. Although various human biological functions may be similar to those of the machines we have created, this does not mean that human beings are machines. Nor does it mean that because certain machines function like human beings, displaying characteristics once thought to be uniquely human, such machines should be considered human beings.

Function cannot indicate essence. Because a machine functions in some ways like a human being does not make it a human being. Because a human being functions like a machine does not make him a machine. If a human being functions like a dog, does it make him a dog?

Philosophers remind us that scientists do not necessarily describe phenomena as they really are but only as they perceive them to be. In the eighteenth century, Immanuel Kant pointed out that we tend to construct ideas or models for understanding reality, but these ideas are only fabrications of our minds which we then impose upon the world. Scientists construct models by which they try to understand reality, but these models are not identical to the reality they attempt to understand. In this regard, William Barrett writes, "The scientist's mind is not a passive mirror that reflects the facts as they are in themselves; the scientist constructs models, which are not found among the things given him in his experience, and proceeds to impose those models upon nature. . . . Here the basic concept of the science, since it is man-made and does not literally copy any single fact in nature, is a product of human artifice and therefore a technical construct as fully as any material piece of apparatus. Here science is technological at its very source." In other words, as the sociologists of science Collins and Pinch have pointed out, science and technology are golems. They are creations of the human mind; they are artifacts that can take on a life of their own.

The understanding of the universe, life, and human beings as machines is one of many models used by contemporary scientists. But it is only a model. Defining a human being is different from defining a chair. The chair does not care how we define it, but a human being has the right to reject or accept how he or she is being understood. In establishing a definition of a human being, I am defining myself. The first test of the validity of such a definition must therefore be its acceptability to myself. Why would any human being wish to be defined as a machine?

Scientists may be able to program machines to look at the world from a human point of view, but they cannot be permitted to impose a machine's view of the world upon us. Indeed, from the machine's point of view, as *R.U.R.* suggests, humans may come to be viewed as imperfect, inefficient, and purposeless beings who are simply impediments to the goals of thinking machines and should therefore be eliminated. Should we accept a view that recommends human obsolescence and extinction? For scientists like Kurzweil and Moravec to do so does not seem to be a problem since, in their view, the inevitable human future is one in which biological humans will extinguish themselves in any case.

Like golems, machines may look and function like human beings, but this need not mean they are humans. Machines may one day have emotions, but theirs need not be human emotions. They may be able to tell jokes, but only other machines and not humans may find them funny.

Like other intelligent beings such as dolphins or even extraterrestrials (should they exist), machines may have intelligence, but that does not mean they have *human* intelligence. The intelligence these beings possess would differ from human intelligence because it derives from different types of experience—from living in a different environment, for example, and from having a different kind of body. A human being's intelligence is shaped by many factors—physiology, environment, individual experience, and the cumulative experience of his own species. Even if these beings have "artificial intelligence" (the term was coined in 1963 by John McCarthy), it is not the same as human intelligence. All of this does not mean that human intelligence is necessarily superior to other forms of intelligence, but it does mean that it is quite different.

If we do not expect our children to be our duplicates or our mimics, why should we expect it of our creations? Why can't intelligent machines, like golems, be considered a species of their own, with their own unique status, rather than a new version of the human species? Why not let human beings remain human and machines remain machines?

Theologians continue to vest human uniqueness in the claims that only human beings have been created in the image of God and that only human beings have human souls. But one need not be a theologian to see that being human is a unique way of what existentialist philosophers call "being-in-the-world." Machines may have the ability to mimic or even to duplicate human intelligence, emotion, locomotion, and speech, but that does not make them human beings. Machines may be programmed with memories, but that does not mean they have human memories. The machine from which I order my airline ticket may sound like a human being, may respond to my questions like a human being, but it is not a human being. Unlike Rava's golem, my automobile may speak to me—like the car on the TV series "Knight Rider"—but that does not make my car a human being.

I can admit that machines have acquired human character-istics without considering them human. I can accept claims that there already are or soon will be machines that surpass human beings in certain functions and abilities, without considering them human. The computer that defeated chess grand master Kasparov is still a machine, and Kasparov is still a human being. I can accept scientific discoveries that inform me that our world is not the center of the universe and still maintain that machines are not human beings and that human existence remains a unique way of being-in-the-world.

A crucial aspect of human being-in-the world is living in a human body. If, as Kurzweil and Moravec predict, we will some-day live "in silico," in computers, or as consciousnesses embod-ied in machines, we shall no longer be human beings. Such a species may evolve from humans, but its members will not be human beings. They will not be capable of experiencing human being-in-the-world, which involves human life in a human body. It is human embodiment that provides the basis for the uniquely human way of being-in-the-world.

For centuries Western philosophers like Plato and Descartes demeaned the human body, preferring instead to extol the soul or the mind. They disparaged the body as the source of moral vice, decrepitude, disease, and death, as representing everything that is finite and corruptible about human beings. Although pretend-ing to represent revolutionary new scientific ideas, contemporary scientists like Kurzweil and Moravec merely offer us an updated version of this long-standing tradition in Western thought. Like Plato more than 2,300 years earlier, they consider the body to be a tomb, a prison from which we must liberate our consciousness and mind. They anticipate a time when we shall be able to divest ourselves of our human bodies and replace them with machines. But when all is said and done, we experience the world as we do precisely because we have human bodies. As embodied human beings, we are creatures of God or of nature. As machines brought into being by humans attempting to design the next evolutionary phase, we would no longer be humans.

In *The Death of the Soul: From Descartes to the Computer*, William Barrett writes: "Our consciousness is embedded in and inseparable from this fleshy envelope that we are. . . . The dreamers of the computer insist that we shall someday be able to build a machine that can take over all the operations of the human mind, and so in effect replace the human person. . . . But in the course of these visions they forget the very plain fact of the human body and its presence in and through consciousness. If that eventual machine were ever to be realized, it would be a curiously disembodied kind of consciousness, for it would be without the sensitivity, intuitions, and pathos of our human flesh and blood. And without those qualities we are less than wise, certainly less than human."

Besides disparaging the human body, much of Western philosophy has sought to deny the reality of human mortality, refusing to accept death as part of human experience. As the early-twentieth-century Jewish thinker Franz Rosenzweig wrote, "Philosophy takes it upon itself to throw off the fear of things earthly, to rob death of its poisonous sting, and Hades of its pestilential breath. . . . And man's terror as he trembles before this sting [death] ever condemns the compassionate lie of philosophy as cruel lying."

From the earliest times, human beings have aspired to be immortal, to be supernatural, to be gods. But as world literature—from the ancient Babylonian *Epic of Gilgamesh*, to the Bible, to the legend of Faust, to the writings of the twentieth-century existentialist philosophers—reminds us, human existence involves mortality rather than divine immortality. Some of our contemporary scientists have not learned this lesson of human history and experience. They look forward to immortal, supernatural existence as the next stage of human evolution, with humans existing as machines or in machines. In his influential *Mind Children: The Future of Robot and Human Intelligence* (1988), Hans Moravec writes, "What awaits us is not oblivion but rather a future which, from our present vantage point, is best described by the words 'post-biological' or even 'supernatural.' It is a world in which the human race has been swept away by a tide of cultural change, usurped by its own artificial progeny." Such a future, should it ever come to pass, would not be a human future. What would happen to human beings like us in such a world? How would our mechanical "children" treat us, relate to us? Would we be placed in zoos as historical relics, as exhibits displaying a now obsolete human past?

Scientists like Moravec have been called "extinctionists," since their plan is to extinguish the human race as we know it. Is the implementation of their vision a new variety of "crimes against humanity"? Ironically, while promising human immortality, Moravec advocates the genocide of the human race.

Kurzweil shares many aspects of Moravec's vision of a post-biological human future, but his is not an idyllic vision. For example, Kurzweil estimates that roughly half the energy of these new human-machines would be devoted to self-protection—defenses against viruses, hostile nanotechnology, and other machines. Certainly if the future species of mechanical humans inherits the vices of their biological forebears, they probably would find themselves neither immortal nor supernatural.

In the 1940s the great mathematician John von Neumann raised the question of whether machines could replicate themselves. Von Neumann was a mathematics professor at Princeton who built MANIAC, the most advanced computer of its day, which enabled the accelerated construction of the hydrogen bomb. He is also credited with inventing "games theory," which became crucial in the development of "war games." Mathematically, von Neumann demonstrated that machines could self-duplicate. Such machines are now being designed and built. Golems, however, cannot replicate themselves. They cannot procreate or reproduce. The authors of the golem legend considered a self-reproducing golem to pose too great a potential danger.

Although some golems in classical Jewish literature (unlike Rava's golem and the golem of Prague) can speak, there is no record of one that can reproduce. The creation (and destruction) of golems had to remain under human control, lest the golem become too powerful and uncontrollable. According to Rabbi Jacob Emden, his ancestor Rabbi Elijah of Helm destroyed the golem he had created because he feared that once it became too powerful the golem might "destroy the world." Granting golems or machines the ability to self-reproduce is viewed as bestowing upon them a power that we may be unable to control. Once they "propagate" their own species, they may consider ours (as in *R.U.R.*) unnecessary, superfluous, and dispensable. If the golems we create exceed our capabilities in power and intelligence, they could readily destroy or enslave us.

In recent versions of the golem legend, golems develop sexual urges and presumably the desire to reproduce. Inevitably this leads them to become destructive—and to their own eventual destruction. Xanthippe, the female golem in Cynthia Ozick's *The Puttermesser Papers*, develops sexual urges and the ability to speak; she evolves beyond her original "programming." Like earlier male members of her species, she goes berserk, and the city she has preserved and defended falls back into disrepair. She sabotages her own achievements. Once her sexual appetite becomes manifest, it grows unrestrained. Every young and handsome man becomes her sexual prey. Is she driven only by a newly acquired libido, or is she, like her creator, motivated by the irrepressible urge to reproduce? Like the golem of Prague, she has gone awry and must be destroyed. And so she is.

In Wegener's 1920 silent film on the golem of Prague, Judah Loew's golem not only evinces a strong sexual interest in Loew's flirtatious daughter Miriam but also—apparently out of jealousy— kills her secret lover, a Czech knight. In Abraham Rothberg's 1970 novel *The Sword of the Golem,* the golem of Prague expresses a strong sexual interest in Rabbi Loew's granddaughter Hava. According to Rothberg, this in itself provides Loew with adequate motivation to deactivate the golem.

Golems are created for a specific purpose. When they exceed their abilities and boundaries, when they attempt to become their own creators or procreators, it is time for them to be destroyed lest they become dangerous and destructive. This vital insight of the golem legend has important meaning for the unfolding of certain current technologies, especially robotics and nanotechnology.

Mechanical golems are now being designed that will have the ability to reproduce, to self-replicate. It is the acquisition of this human capability that poses the greatest danger to the human creators of golemic machines. For this reason, strong opposition to the creation of intelligent machines with the ability to self-replicate has come from various quarters, some unexpected—like the computer genius Bill Joy.

Joy could never be considered a Luddite. He has been a key figure in the development of groundbreaking computer software. He invented the basic computer language called Java and was one of the founders and the chief scientist of Sun Microsystems. In April 2000, Joy offered his thoughts about technology in the new millennium in an article in the leading technological journal *Wired*. To the shock of most readers, Joy insisted that scientists pull back from genetic research, nanotechnology, and advanced robotics. It was as if the president of the American Medical Association had demanded that his colleagues relinquish their research in pharmacology and medical technology. Joy wrote that after an extended conversation with Ray Kurzweil, he became convinced that accelerating developments in robotics, computer science, and nanotechnology posed a clear and present danger to human life on earth as we know it.

Today nanotechnology is a "hot" field of research. Nanotechnology is the art and science of manipulating and rearranging individual atoms and molecules to create materials, devices, and systems. It deals with infinitesimally small things. It is the willful manipulation of matter on the atomic level, to turn things into other things. As the Nobel Prize–winner Horst Stormer has observed, "Nanotechnology has given us the tools . . . to play with the ultimate toy box of nature—atoms and molecules. Everything is made from it. . . . The possibilities to create new things appear limitless."

Nanotechnology measures things by nanometers; a nanometer is 1/80,000 the diameter of a human hair. Nanotechnology promises to produce stronger, lighter, and more flexible materials, to transform polluted and dangerous materials into useful materials. Nanorobots might be injected into the human body to diagnose disease and then to repair damaged cells. Nanotechnology is already used on clothing to repel rain and prevent stains. It has the potential to transform many of our industries and businesses as well as the delivery of health care. It may also transform the art and practice of warfare by producing generations of new and horrific weapons.

Central to the successful application of nanotechnology will be the ability of entities like nanorobots to self-duplicate. Proponents of nanotechnology envision a world free from material want because anyone will be able to build anything. Opponents fear that these devices will be uncontrollable and destroy anything in their path.

What specifically troubled Bill Joy was the prospect of biotechnology, computers, and nanotechnology being able to create entities that could self-replicate—specifically, entities that are harmful, dangerous, and lethal. Not only invasive and destructive computer viruses might be duplicated; potentially devastating biological viruses could be designed and duplicated that could threaten human life. Rapidly duplicating swarms of nanorobots, as portrayed in Michael Crichton's novel *Prey,* could place all life on earth in danger of extinction as they "feed" on every available organic and nonorganic matter, including human beings. This threat posed by nanotechnology is known as "the gray goo problem"—the remote possibility that self-replicating nanorobots could transform everything on our planet into some form of slime. Scientists assure us that they can control nanotechnology, but Crichton warns that "it is always possible that we will not establish controls. Or that someone will manage to create artificial, self-reproducing organisms far sooner than anyone expected. If so, it is difficult to anticipate what the consequences might be."

Joy would agree with Crichton that the potential dangers of self-replicating nanorobots are too horrible to contemplate. Nuclear weapons, Joy reminds us, were developed in secret and placed under strict government control. Current and expected developments in the creation of self-replicating machines, on the other hand, are happening "in the open," often under the aegis of revenue-driven corporations. The development of atomic weapons in the 1940s required large, expensive facilities and rare materials, but current technologies are readily and cheaply available. This new technology could empower dangerous individuals with powers they ought not to have.

Reflecting on the difference between the development of the atomic bomb and current developments in nanotechnology, Joy writes, "Robots . . . share a dangerous amplifying factor: They can self-replicate. A bomb is blown up only once—but one robot can become many, and quickly get out of control. . . . Uncontrolled self-replication in these newer technologies runs . . . a risk of substantial danger in the physical world."

Joy sees an "intelligent" robot, with intellectual capabilities equal to or exceeding those of human beings, available by the year 2030. "Once an intelligent robot exists," he notes, "it is only a small step to a robot species—to an intelligent robot that can make evolved copies of itself." With Kurzweil and Moravec in mind, Joy adds, "It seems to me far more likely that a robotic existence would not be like a human in any sense that we understand, that robots would in no sense be our children [as Moravec suggests], that on the path our humanity may well be lost."

Joy fears for the future. "This is the first moment in the history of our planet when any species, by its own voluntary actions, has become a danger to itself—as well as to vast numbers of others. . . . An immediate consequence of the Faustian bargain in obtaining the great power of nanotechnology is that we run a grave risk—the risk that we might destroy the biosphere on which all life depends." If we lose control, if we make mistakes, if we botch experiments, the results could be devastating and apocalyptic.

The risks of catastrophe are substantial. Joy quotes one study that placed the risk of human extinction from our own technologies at no less than 30 percent. Kurzweil's view is that we have "a better than even chance of making it through." But Kurzweil adds that he is "always accused of being an optimist." With these odds and with the development of technologies of machine replication, would a life insurance company write a policy on the human race?

The last lines of Crichton's *Prey*, about nanotechnology, contain a dialogue:

> "They didn't understand what they were doing."
> "I'm afraid that will be on the tombstone of the human race."
> "I hope it's not."
> "We might get lucky."
> "But what if we don't?"

It is not without wisdom that the golem legend prohibits the creation of golems that can procreate. The legend insists that artificial creatures should always remain under our close control, and demands that we always retain the ability to destroy them if they threaten to become dangerous or deadly. Similarly, the Foresight Institute, founded by Eric Dexler, a pioneer in nanotechnology, in its 1999 Guidelines on Molecular Nanotechnology forbids the creation of nanorobots capable of "replication in a natural controlled environment," and offers various ways of enforcing such guidelines.

Our coexistence with the mechanical golems that dwell among us has had a profound influence, not only on the physical and material aspects of our daily lives but on our spiritual lives as well. It has influenced our self-perceptions and our understanding of the world. The long-standing tendency to think of the world, and ourselves, as machines has affected us enormously. Yet the more we think of ourselves as machines, the more machinelike we will become—and the more we will surrender our humanity, our souls, our freedom—our spiritual nature—to a machinelike existence.

The more we come to depend on machines, the more we lose control over them and become their servants, controlled by their directions as well as their idiosyncrasies—as anyone who has used a personal computer knows only too well. As machines become more and more complex, we grow increasingly dependent upon and under the control of those who design them and know how they operate. As in *R.U.R.*, machines are assuming so many human tasks that human beings may readily become obsolete, superfluous, and without a clear sense of purpose. If the computer-science-driven categories of function and efficiency become the criteria for evaluating the nature and quality of our actions and our lives, we may indeed be rendered obsolete, surpassed and replaced by the more efficient functioning of machines. The value of human life will become secondary to the value of the technological artifacts we create. Machines may become more like humans, and humans more like machines. But the greater danger is that machines will strip us of our uniquely human manner of being-in-the-world.

In the final analysis, machines are golems and we are human beings—neither golems nor machines. The possible evolution of machines as postbiological humans, the growing erosion of human autonomy and its surrender to machines, violates human nature, freedom, creativity, and the uniquely physical and spiritual manner of human being-in-the-world. As Joy says, the machines of the present and future are not our progeny, as Hans Moravec suggests. Nor can the artifacts created to serve us be permitted to become our masters. The mechanical golems we create may dwell among us, but they are not us, nor are we them. To consider ourselves machines, and to consider machines to be human beings, opens the door to the physical and spiritual suicide of the human race. This is precisely why the golem legend emphatically insists that we closely control the artificial beings we have created.

Developments in robotics, artificial intelligence, computer science, nanotechnology, and bionics may provide us with a future that can be catastrophic or utopian. The challenge for us is to exercise and assert our freedom, before choices are made for us regarding who we want to be and how we want to live in the rapidly approaching, increasingly mechanized future, before we surrender control to the artifacts we have created. To meet this challenge we need not only intelligence but wisdom, including the wisdom of Scripture that reminds us, "What shall it profit a person if he gains the whole world, but loses his soul?"

Chapter Nine

Corporate Golems: The Supreme Court Creates an Artificial Person

When we think of the creation of artificial persons, we tend to conjure up images of a mad scientist like Dr. Frankenstein, a master of the mystical arts like Judah Loew of Prague, or a team of cutting-edge engineers. Yet in law offices and courts throughout the United States, artificial persons are being created every day. They are known as corporations.

In his classic eighteenth-century work *Commentaries on the Laws of England,* William Blackstone summed up earlier traditions in British common law when he distinguished between two types of "legal persons": natural and artificial. In Blackstone's words, "natural persons" are "such as the God of nature formed us," while "artificial persons" are "such created and devised by human beings for purposes of society and government." A human being is a "natural person"; a corporation is a prime example of an "artificial person." Like golems, corporations are entities created by human beings to serve human needs and to labor on their behalf. In other words, a corporation may be considered one of the various kinds of golems that currently populate our world. According to Blackstone, "a corporation [cannot] be excommunicated for it has no soul." Like a golem, a corporation is a body—a corpus, without a human soul. Like a golem, a corporation is impervious to physical pain. No sword or bullet can harm it. No prison can contain it.

Like golemic power, corporate power needs to be held in check lest it become uncontrollable, lest corporate golems become corporate Frankensteinian monsters. The problem that confronted other modern creators of golems now confronts public policymakers and society at large with regard to large commercial corporations: Can we control the artificial beings we have created to help us before they harm or destroy us, and if so, how?

In the case of many of the golems that currently populate our world—some of which have been described above—there is still time to control them. For the most part they have not wreaked the havoc that some have predicted they would. Robots, for example, have neither overwhelmed nor replaced us. We have not become their servants, nor they our masters. Yet corporations have become masters of many of us and of many facets of our lives. Large multinational corporations already have caused irreparable harm to the lives, health, and fortunes of many innocent people. Left to their own devices, their power permitted to expand without controls, they will inevitably continue to threaten our lives, our health, our economic well-being, and our liberties.

157

Like golems, corporations can do many useful and beneficial things. In our time they have helped improve the lives of millions. They have provided jobs and livelihoods for countless people. They have produced unprecedented wealth and prosperity. They have helped to build towns, cities, and transportation systems. They have developed new lifesaving pharmaceuticals and introduced many affordable conveniences. They provide more of the necessities and even the luxuries of life than anyone could once have imagined. They have helped build armies and navies to defend us. Indeed, corporations have been instrumental in the building of nations, including our own. Without them it is doubtful that America would have emerged as a global superpower in the twentieth century—as the world's only superpower in the twenty-first century. Millions upon millions of Americans owe much of their wealth, prosperity, careers, and comforts to the ingenuity, productivity, and efficiency of corporations.

But, like some golems, corporations can run amok. They can expand in size and power and wreak havoc and destruction. They can manipulate governments, corrupt politicians, destroy careers, deplete the wealth of their employees and investors, cause environmental damage, avoid taxation, and commit crimes and heinous deeds—often with impunity. At the onset of the twenty-first century, corporate scandals sent tremors through the American and world economies. A new word entered the English language: "Enronization."

The U.S. Supreme Court Justice Louis Brandeis (1856–1941) was a direct descendant of Rabbi Judah Loew of Prague. It may not be coincidental that Brandeis, like his ancestor, advocated both the creation as well as the need to control the power of artificial beings. In his opinion in the 1933 Supreme Court case of

Liggett v. Lee, Brandeis traced the history of the development of corporations and pleaded for their control.

As we have seen, modern versions of the golem legend maintain that golems were created to serve human needs, to act on behalf of the public good and benefit their creators and owners. Later on, golems were portrayed as being potentially dangerous, though still under the control of their creators. As the golem legend developed further—probably under the influence of tales like *Frankenstein*—golems began to be portrayed as malevolent, out-of-control monsters. In his survey of the development of the status of the corporation until his own time, Brandeis similarly saw corporations as initially beneficent beings, created for the public good, that later became inordinately powerful and therefore potentially dangerous creatures. Subsequently, just as the malevolent Frankenstein monster eluded the control of its creator, corporations not only avoided oversight by the states that created them but began to exercise untoward control over governments and their citizens.

For Brandeis, the expansion of corporate power led to the inception of a new "feudal system" where masses of people became the serfs of large corporate conglomerates, where a controlling share of national wealth was accumulated by a comparatively small number of enormous corporations, and where there amassed "such a concentration of economic power that so-called private corporations are sometimes able to dominate the state." Such corporate entities Brandeis found guilty of "thwarting American ideals [such as] . . . equality of opportunity." Large corporations were not golems, he maintained but "the Frankenstein monster which states have created by their corporate laws." If this was true in Brandeis's time, how much more true is it today when most corporate power is vested in large, publicly held corporations? It is to such enormous commercial corporate entities that most of the following discussion refers. It is not the 99.6 percent of American corporations that have less than $10 million in capitalization or the nonprofit corporations that are worrisome. It is rather the Fortune 1000 corporations that now control about 70 percent of the American economy.

The transition of corporations from golems into "the Frankenstein monster" seems to have begun during the American Civil War. It was then that corporate entities rapidly expanded by gorging themselves on the gains of wartime profiteering. At the same time American industry began to experience exponential growth. As the railway and steel industries mushroomed after the Civil War, corporate "robber barons" tried to control as many people, as much wealth, and as much influence over the government as they possibly could.

On November 21, 1864, as the Civil War was nearing its end, Abraham Lincoln sent a letter to his friend Col. William F. Elkins. Reflecting on the enormous loss of life the nation had suffered during the war, Lincoln wrote, "We may congratulate ourselves that this cruel war is nearing its end. It has cost a vast amount of treasure and blood. The best blood of the flower of American youth has been freely offered upon our country's altar that the nation might live. It has indeed been a trying hour for the Republic; but I see in the near future a crisis approaching that unnerves me and causes me to tremble for the safety of my country." What did Lincoln fear? Just as the nation stood poised to conclude the conflict that almost destroyed the country forever, Lincoln saw a crisis looming of even greater danger to the future of the United States than the Civil War.

Lincoln continued and explained, "As a result of the war, corporations have been enthroned and an era of corruption in high places will follow, and the money power of the country will endeavor to prolong its reign by working upon the prejudices of the people until all wealth is aggregated in a few hands and the Republic is destroyed. I feel at this moment more anxiety than ever before, even in the midst of war. God grant that my suspicions may prove groundless."

Rather than being groundless, Lincoln's fears were prescient. In the decades after his assassination, corporate corruption, especially in the railroad industry, spread like an aggressive cancer. While law enforcement pursued train robbers, the leaders of the railway industry sought to appropriate land, avoid paying taxes, compromise government officials, exploit immigrant labor, and subvert the judicial process to their own advantage.

The railroad corporations yearned to be liberated from the control and authority of the individual states that initially had chartered them—that is, from their own creators. They wanted certain rights and privileges that would further their corporate goals and protect them from government oversight. At the same time they wanted to be free from prosecution and liability for their criminal activities. What they sought was a legal mechanism that would help them get what they desired.

As artificial persons, corporations did not enjoy the protections of constitutional rights granted to natural persons—to human beings. Since the Constitution does not mention corporations, the railroads were subject to state law, with none of the legal protections afforded by federal constitutional law. On the other hand, as artificial persons, corporations were considered incapable of committing certain types of crimes or of receiving certain kinds of punishment, such as imprisonment. So the railroad corporations saw that their ideal situation would be to retain the status of artificial persons in so far as criminal liability and responsibility were concerned, while being granted some of the rights and protections of the Constitution. For this to occur, the Supreme Court would have to declare that the corporation was a natural person—enjoying certain constitutional rights and privileges—yet as an artificial person remaining free from criminal liability. Would the Court agree to codify such a bizarre precedent, making it the law of the land? That was the question that arose in the 1886 case of *County of Santa Clara v. The Southern Pacific Railroad.*

Like many of the "railroad cases" of that era, the dispute was the same: a local government had tried to collect certain taxes from the railroad company, which refused to pay. Apparently at issue here was whether the railroad was obliged to pay a county tax of about one-tenth of 1 percent. But the real issue was whether corporations could enjoy the status of natural persons, specifically the constitutional protections that due process of law guaranteed natural persons under the Fourteenth Amendment.

Even before arguments were heard in this case, Chief Justice Morrison Waite apparently spoke out on this issue. Waite had ascended to the nation's highest judicial position though he had never before served as a judge. He had enjoyed a distinguished career as an attorney defending railroads and other large corporations. Waite was reported to have declared that "The defendant corporations are persons within the intent of the clause of section one of the Fourteenth Amendment to the Constitution of the United States, which forbids a state to deny any person within its jurisdiction the equal protection of the laws."

What this meant, in effect, was that an "artificial person" such as a corporation, enjoyed the same *legal* status as a "natural person." Waite's statement effectively eliminated the distinction between artificial and natural persons before the law. The Court, he hinted, would treat corporations as if they were natural persons—human beings. To complicate the matter further, in the *Santa Clara* case the Court seemed only to grant artificial persons the rights and protections of law as they applied to natural persons, without addressing the vital issue of whether they also had the obligations and responsibilities of natural persons. It seemed as if the railroad corporations had gotten exactly what they had hoped for.

Yet the text of the decision in the *Santa Clara* case makes no reference to Chief Justice Waite's statement. It is recorded only as a headnote by the court reporter, J. C. Bancroft Davis. Before working for the Court, Davis had been a railroad president. Recently a note from Waite to Davis regarding this case was found in the National Archives. It says, "We avoided meeting the Constitutional question in the decision." In other words, neither Waite nor the Court in *Santa Clara* had in fact bestowed upon artificial persons the rights of natural persons. A precedent for applying the Fourteenth Amendment to corporations was never really made in the case. Chief Justice Waite probably never said what he is alleged to have said about extending Fourteenth Amendment rights to corporations. The court reporter's headnote technically had no precedential status. Indeed, by the time it was published, the chief justice had died. Nonetheless this headnote was treated as if it had been part of the decision, and ever since then the *Santa Clara* case has been considered a precedent for granting corporations the legal status of natural persons, such as the protections of the Fourteenth Amendment.

Accepting the precedent allegedly established in the *Santa Clara* case, justices subsequently struck down laws enacted to protect people from corporate abuse. For example, the 1905 case of *Lochner v. New York* reaffirmed the Fourteenth Amendment rights of corporations and invalidated approximately two hundred regulations related to government oversight of corporate behavior. *Lochner* later served as a precedent for similar cases where corporations sought to limit government oversight. Now, rather than the government limiting the scope of corporate activity, corporations—with the help of the Court—were limiting the government's control over themselves.

While the Supreme Court in the late nineteenth and early twentieth centuries continuously reaffirmed the rights of corporations to equal protection and due process as guaranteed by the Fourteenth Amendment, it simultaneously denied many of those same rights to human beings, including women and racial minorities. In the first 45 years after the ratification of the Fourteenth Amendment, which was adapted for the express purpose of granting rights to African-American men in the wake of the Civil War, 604 cases relating to the Fourteenth Amendment were heard by the Supreme Court; only 28 concerned the rights of African-Americans while 312 involved corporate rights. More often than not, the rights of corporations were upheld while those of African Americans were not. Ironically, in 1868, the year the Fourteenth Amendment became law, the Supreme Court in *Paul v. Virginia* had ruled that under Article 4, Section 2 of the Constitution, corporations were not citizens and therefore could not enjoy the constitutional rights and protections of citizens.

While dispensing with the long-term legal distinction between artificial and natural persons, *Santa Clara* in effect granted corporations greater rights than those of natural persons.

Natural persons still remained responsible and liable for committing crimes that corporations could often commit with impunity. Precisely because large corporations wielded greater power and influence than individual citizens, and could hire better legal representation than most individuals, large corporations could better assert and defend their legal rights than could the average citizen.

The implications of *Santa Clara* were not lost on President Grover Cleveland. Two years after that case, addressing a joint session of Congress and apparently with *Santa Clara* in mind, Cleveland observed, "As we view the achievements of aggregated capital, we discover the existence of trusts, combinations, and monopolies, where the citizen is struggling far in the rear or is trampled to death beneath an iron heel. Corporations, which should be the carefully restrained creatures of the law and the servants of the people, are fast becoming the people's masters."

In Supreme Court cases after *Santa Clara*, corporations won additional constitutional rights to those protected by the Fourteenth Amendment. For example, in the 1906 case of *Henkel v. Henkel*, corporations were granted protection under the Fourth Amendment against "search and seizure" of corporate records. Other Court cases have granted corporations rights, privileges, and protections under the First, Sixth, and Seventh Amendments.

As the rights of corporations expanded, so did their power, influence, control, and corruption. By the beginning of the twentieth century, Lincoln's fears were rapidly becoming reality. Monopolistic corporate trusts had grown so large, and become so powerful and wealthy, that they controlled entire industries vital to the American economy—oil, steel, banking, and railroads. They had enormous influence on the American economy, on the political and judicial systems, on international affairs, and on the lives and liberties of American citizens.

Early in his presidency, Theodore Roosevelt identified the "dark power" of corporations, especially the trusts. Until he did so, the 1896 Sherman Antitrust Act had no real effect on regulating American industry. Contrary to the conditions of his day, Roosevelt insisted that the government exercise control, regulation, and oversight of corporations. As he put it, "There can be no effective control of corporations while their political activity remains. To put an end to it will be neither a short nor an easy task, but it can be done." Yet in fact it was never accomplished.

By the time Theodore Roosevelt's distant cousin Franklin became president, corporations not only threatened to control governments, including that of the United States, but to become like sovereign states of their own. In 1921, President Warren G. Harding had codified the view that there should be "less government in business and more business in government." But in his acceptance speech for a second-term nomination at the 1936 Democratic National Convention, Franklin Roosevelt charged, "Concentration of economic power in all-embracing corporations . . . represents private enterprise become a kind of private government which is a power unto itself—a regimentation of other people's money and other people's lives."

By 1961 the outgoing President Dwight Eisenhower was warning the American people of the dangers of the "military-industrial complex," in which the state, military, and corporate interests had meshed with potentially devastating consequences. In Eisenhower's words, "This conjunction of an immense military establishment and a larger arms industry is new in the American experience. The total influence—economic, political, even spiritual—is felt in every city, every state house, every office of the Federal government."

Many historians believe that Eisenhower's warning to America, delivered at the end of his presidency, reflected frightening comparisons in his mind between the direction he saw America taking in the early 1960s and what he had learned about the Nazi Third Reich during the liberation of Europe at the end of World War II.

The issue of corporate moral responsibility had been raised in a powerful and poignant way just after World War II in the trials of Nazi war criminals. The evidence presented at these trials demonstrated the danger of corporate power when left unchecked, and in particular the relationship between the Third Reich and the international conglomerate I. G. Farben.

Because Auschwitz has become a symbol of the Holocaust, it is important to note that the Nazi war crimes trials concluded that "Auschwitz was financed and owned by Farben. The use of concentration camp labor and forced foreign workers at Auschwitz, with the initiative displayed by the officials of Farben in the procurement and utilization of such labor, is a crime against humanity." In other words, Auschwitz was merely a subsidiary in the huge corporate empire of I. G. Farben.

Like I. G. Farben, many other leading German corporations, including BMW, Daimler-Benz, Krupp, Siemens, and their subsidiaries, aggressively used "slave labor" during World War II, mostly concentration camp inmates. No German corporation was required by the state to use these laborers. Rather, these corporations used their influence and resources to gain permission for the use of inmates as slave laborers. Consequently German corporations invested considerable sums of money in the construction of facilities at concentration camps and forced-labor camps that would meet their needs and those of their "partners," the SS. Millions of reichsmarks were paid to the SS for the "privilege" of using slave laborers that the SS would supply. Farben's investment in Auschwitz alone exceeded one billion reichsmarks. Never before in human history had an artificial person, a corporation, had so many human beings as its slaves. The servant became the master, with catastrophic results.

Nazi government policy was to murder all members of certain targeted groups, especially Jews. But German business disagreed—not for humanitarian reasons but purely for profit-driven motives. Given the severe labor shortage in Germany during the war, workers were a valuable commodity. The concentration camps could supply an almost limitless supply of skilled and unskilled laborers. If those inmates who could work were used (while others were murdered in the gas chambers), the labor shortage could be alleviated. As a Nazi directive put it, in order to ensure maximum profitability, "All inmates must be fed, sheltered, and treated in such a way as to exploit them to the highest possible extent at the lowest conceivable degree of expenditure." In other words, they were starved and worked to death, then replaced by new workers. In more than fifteen hundred concentration camps and subcamps in the early 1940s, more than half a million people were "leased" out by the SS as slave laborers to hundreds of German corporations.

As Joseph Borkin wrote in his 1978 book *The Crime and Punishment of I. G. Farben,* "The construction of I. G. Auschwitz has assured I.G. [Farben] a unique place in business history. By adopting the theory and practice of Nazi morality, it was able to depart from the conventional economics of slavery in which slaves are treated as capital equipment to be maintained and serviced for optimum use and depreciated over a normal life span. Instead, I.G. reduced slave labor to a consumable raw material, a human ore from which the mineral of life was systematically extracted. When no usable energy remained, the living dross was shipped to the gassing chambers and cremation furnaces of the extermination center at [Auschwitz-]Birkenau, where the SS recycled it into the German war economy—gold teeth for the Reichbank, hair for mattresses, and fat for soap. Even the moans of the doomed became a work incentive, exhorting the remaining inmates to greater effort."

Auschwitz was not one camp but a complex of camps. Auschwitz II, or Auschwitz-Birkenau, existed for the purpose of gassing people to death and cremating their remains. Those who could not work upon arrival, and those who would no longer provide productive labor, ended up there. Auschwitz-Buna and Auschwitz-Monowitz were slave labor camps. One of the goals of these I. G. Farben–owned subsidiaries was the production of synthetic oil and fuel to be used in the German war effort. Ironically, Farben's investment in these camps—at the cost of thousands of human lives—did not pay off. Only a modest stream of fuel and no rubber were produced there.

Soon after the war, Eisenhower ordered an investigation into Farben's role in the German war effort. The report of this investigation concluded that the company had been indispensable to German's ability to make war. Without Farben, Hitler could neither have launched the war nor come so close to victory. It was further discovered that Farben had produced terrible chemical weapons that could have won the war. They were not used

because Hitler incorrectly believed that the Americans would re-
taliate with comparable weapons. Based on the report of Farben's
participation in the German war effort and in the operation of
concentration camps, Eisenhower ordered the company's dismantling as "one means of assuring world peace."

As punishment for its enthusiastic participation in Nazi war
crimes, Farben was broken up after World War II. Yet today many
of its former divisions are independent corporations, each larger
than I. G. Farben had been during the war. By 1977, former Farben subsidiaries such as BASF and Bayer were larger than Farben had been at its zenith. By 1951 none of the corporate leaders
convicted of war crimes committed during World War II was still
imprisoned.

Eisenhower may have also known that even during World
War II, German and American corporations continued to do
business with one another that damaged the American war effort; that American shareholders in American companies received dividends from profits by American companies that were
trading with the enemy in wartime; and that these companies
had interlocking and mutual interests with German companies
during the war. This was true of such leading American corporations as IBM, Ford, and Standard Oil. DuPont continued to
own 6 percent of Farben common stock during the war. Standard Oil, in partnership with Farben, owned the patents on the
manufacturing process for synthetic oil and rubber, a major goal
of the Farben-owned facilities at Auschwitz. U.S. attempts to destroy the Farben–Standard Oil international corporate cartel
were stifled by American government officials both during and
after the war. Despite flagrant violations of the Trading with the
Enemies Act by these corporations, little action was taken
against them during or after the war. Indeed, many of these

relationships between American and German corporations extended decades after World War II.

If individual American citizens—natural persons—had given aid and comfort to the enemy in a time of war, we would expect nothing less for them than the harshest punishment. But as Blackstone had pointed out, corporations cannot be given corporal punishment, nor can they be imprisoned. Perhaps this is why huge international corporate cartels thought they could trade with the enemy during wartime without consequences. In 1942, while millions of Americans had to endure long lines at gas stations and use their gas coupons cautiously, Standard Oil of New Jersey was shipping oil to the enemy through neutral Switzerland. Meanwhile, Ford trucks were being built for German occupation forces in France.

The story of I. G. Farben demonstrates how far the power of corporate conglomerates can extend if left unchecked. Farben was an artificial person, a body without a soul, but with demonic prowess and power. As international corporate cartels continue to expand their power and influence, Farben-like and Frankenstein-like beings threaten to surface. The warnings of Presidents Lincoln, Theodore Roosevelt, and Eisenhower about the dangers of uncontrolled corporate power have become ever more relevant and poignant.

The confluence of new technological capabilities and unbridled corporate greed can lead to the commodification and dehumanization of human life. The health-care industry is a good example. Dedicated to the preservation of life and health, scientists, physicians, and drug manufacturers provide new pharmaceuticals that save, extend, and enhance life. Yet, like other corporate industries, they can succumb to the temptation of valuing products more than the sick and dying.

With the end of the cold war, the spymaster novelist John Le Carré has turned his attention to current threats to individual human dignity and freedom. In *The Constant Gardener* (2001), he locates them in the practices of international corporate conglomerates. Here Le Carré focuses on the pharmaceutical industry. He portrays a world of corporate deception, manipulation, and unbridled greed. He depicts corporate-instigated character assassination, manipulation of governments, and even murder. He describes experiments using experimental drugs on unwitting human subjects, especially in Africa, where international sanctions against such activities seem not to apply. In an author's note, he writes, "I can tell you this. As my journey through the pharmaceutical jungle progressed, I came to realize that by comparison with the reality, my story was as tame as a holiday postcard." Similarly, the eminent biologist Erwin Chargoff describes the sometimes predatory and exploitative practices of the biotech and pharmaceutical industries to secure monopolistic patents on human genes and tissues as "an Auschwitz in which valuable enzymes, hormones, and so on will be extracted instead of gold teeth." A case in point of the many that might be cited is that of *Moore v. Regents of the University of California*.

In the late 1970s, John Moore was diagnosed with and treated for a form of leukemia at UCLA Medical Center. In the course of his treatments, extensive amounts of blood, bone marrow and bodily tissue were withdrawn from his body. Without his knowledge, these substances were used by his physicians and others for research purposes aimed at developing future medical treatments for other human beings. Also concealed from him was the expectation of his physician and his physician's colleagues of benefiting financially from the products of their research with his cells. His physician used his privileged relationship with Moore for exclusive access to substances from Moore's body. The extraction of these substances was not always directly related to Moore's ongoing treatment but was part of this research activity.

Based upon the research of Moore's physician's group, a patent on the cell line developed from Moore's cells was applied for and was granted in 1984. Moore's rare cell line would be used to produce certain pharmaceuticals. Moore's physician and his partners, together with the UCLA Medical Center (the owners of the patent), sold rights for the commercial development of pharmaceuticals derived from Moore's cell line for a seven figure amount. It had already been estimated that the sale of such pharmaceuticals would bring a multi-billion-dollar yield within seven years. When Moore learned of all this, he brought suit for malpractice and property theft. He argued that he deserved some part of the funds generated by the use of his own bodily substances, especially those taken for research and not for therapeutic purposes, and without his knowledge of their eventual use. The case went to the Supreme Court of California, and Moore lost. His comment: "I was harvested."

The court acknowledged that the physicians had lied by telling Moore that no financial or commercial value could be derived from his blood and bodily substances. The court understood that the physicians had consistently concealed their plans for economic gain from Moore. Nonetheless the court noted that the "defendants who allegedly obtained the cells from the plaintiff by improper means, [can] retain and exploit the full economic value of their ill-gotten gains free of liability."

Citing *Diamond v. Chakrabarty* as a precedent, and expressing concern that to "restrict access to the necessary raw materials" would hinder the progress of biomedical research, the court's majority ruled against Moore. A minority, however, stated that because of the "defendants' moral shortcomings, duplicity, and greed," they should be compelled "to disgorge portion of their ill-gotten gains to the individual whose body was invaded and exploited and without whom such profits would not have been possible." In his dissent, Justice Mosk wrote that the "specter [of the abuse of the human body for economic exploitation] haunts the laboratories and boardrooms of today's biotechnological research-industrial complex. It arises whenever scientists or industrialists claim, as defendants claim here, the right to appropriate and exploit a patient's tissue for their sole economic benefit—the right, in other words, to freely mine and harvest valuable physical properties of the patient's body. Research with human cells that results in significant economic gain for the researcher and no gain for the patient offends the traditional mores of our society in a manner impossible to quantify. Such research tends to treat the human body as a commodity, as a means to an end."

What is now being called "biopiracy" and "biocommerce" requires legal and ethical limitations. When Jonas Salk invented the polio vaccine, he refused to apply for a patent on it because he considered it more a product of nature than of his own ingenuity. "Could we patent the warmth of the sun?" he asked. In the 1990s the American Medical Association amended its code of ethics to forbid doctors from patenting medical procedures because it found that these patents compromised patient care. Meanwhile the justices in the Moore case denied him the right to his own bodily tissues because they believed that otherwise they would "destroy the economic incentive to conduct important medical research." What seems to have eluded the court's reasoning is that monopolistic patents on human genes and tissues can impede as well as encourage the development of truly useful diagnostics and therapeutics. Certain patents on tissues, genes, cells that cause disease, or on medical procedures may serve to restrict the use and development of new and improved scientific discoveries by other researchers aimed at therapeutic treatment or diagnosis of various diseases.

While biotech and pharmaceutical companies should not be denied the opportunity to make reasonable profits from their research, labor, and ingenuity, and while their development of life-saving and health-enhancing drugs is to be welcomed, the dehumanization and exploitation of human subjects in the securing of monopolistic patents and licenses cannot be morally defended. Companies and individuals awarded such patent and licensing rights could readily withhold lifesaving drugs that they consider not adequately profitable. They could make patented drugs available only at astronomical prices and collect huge, inflated profits. One way of dealing with this situation would be to limit both the exclusivity of these patents and the profits realized from them. Ironically, the basic research that has led to many of these patents was initially undertaken with government grants— with taxpayer dollars. For example, taxpayers who financed the $3 billion cost of the U.S. government's Human Genome Project may be excluded from partaking of its benefits. Of the fifty best-selling drugs, forty-eight benefited from federal research money in their developmental or testing phases. At the very least, taxpayers should get something in return for their dollars, like lower prices for drugs and diagnostics developed with their own money. One way of regulating what has been called the "Biotech Gold Rush" would be to institute a "reasonable price" criterion on pharmaceuticals, especially on those drugs, genetic tests, and medical procedures that have been developed with public funds.

In *Body Bazaar* (2001), Lori Andrews and Dorothy Nelkin wrote, "As biomedical research becomes even more closely tied to commercial goals, the encroaching market is triggering a growing sense of disillusionment and mistrust. For the application of commercial practices to the human body are increasingly challenging individual and cultural values, encouraging exploitation through the collection and use of tissue, and turning tissue and potentially people into the marketable products of a body bazaar."

A world governed by gigantic international corporations, described in the 1975 futuristic film *Rollerball*, is rapidly moving from fiction to reality. In such a world the power of nations may be challenged by the power of huge multinational corporations. Governments may surrender or compromise their sovereignty to large corporate entities. When one considers that already today, of the hundred largest economies in the world, fifty-one are corporations while forty-nine are countries, the corporate domination of the natural and human resources of the world is no longer a futuristic fantasy. The remake of *Rollerball* in 2002 did not envision a world controlled by corporations in the distant future. The film is set in 2005. From what has already been said about corporate power in the twentieth century, one would think that people in the twenty-first century would have learned that corporate power is too dangerous to be left uncontrolled. Recent corporate scandals in the United States should have reinforced that message.

To satisfy corporate goals, some large transnational corporations continue to break laws and ignore moral principles with impunity. They violate human and civil rights, damage the environment, exploit child labor, and manipulate both national laws and international treaties in their favor. While the welfare rolls of the poor have been reduced, government-funded corporate bailouts, subsidies, and tax incentives have increased. While tax evasion among individual citizens is met with harsh penalties, corporate offshore tax havens aimed at tax evasion and avoidance abound. (The Cayman Islands is populated by more corporations than people.) Dummy corporations are created to further extend fiscal and tax deceptions and fraud. Many fabulously profitable corporations pay a much lower tax rate than most of their own employees. According to the General Accounting Office (GAO), almost a third of large U.S. corporations (with assets over $250 million dollars) paid no income tax between 1989 and 1995. According to the GAO, in those same years a majority of corporations, both U.S. and foreign-controlled, paid no U.S. income tax. In four of the five years before its collapse in 2002, Enron paid no federal taxes.

Although petty criminals are imprisoned for long sentences, crimes committed by corporations and by their agents and executives often go unpunished—or they are punished with lenient fines and sentences for substantial illegal activities. Corporations pay enormous settlements while admitting no responsibility or liability for damage. Although robberies and burglaries cost the American people about $4 billion a year, it has been estimated that corporate crimes and fraud cost us hundreds of billions annually.

In the wake of the Enron scandal, a call for corporate reform has again been heard throughout the land. Once more a U.S. president has called for action against corporate crimes and malfeasance. Again new laws and regulations, like the Corporate

Responsibility Act, have been enacted. Yet the untold damage to people's lives, careers, hopes, and expectations, and to the national economy, caused by corporate malfeasance remains largely unaddressed. Even if corporations and their executives were punished, would the punishment fit the crime? Could the punishment undo the damage caused by such crimes? Is the threat of fines an adequate deterrent, especially when paying them is calculated into the corporate balance sheet as a cost-effective manner of doing business? If the money lost is no longer there, how is financial restitution possible? Who should be held responsible for corporate crimes?

The stockholders who invest in corporations—who own them—have been increasingly marginalized in their management, especially in large corporations. They have little say about how and by whom the corporations are run. As they have little power, they also have—by law—limited responsibility and liability. Stockholders have no financial liability beyond their own investment. The value of their stock may rise or fall, but beyond that they have no personal liability for corporate malfeasance. Although they own part of the corporation, they have no personal culpability for corporate crimes, misdeeds, or bad management. As Justice Brandeis noted, once ownership becomes detached from management, corruption and abuse is likely to follow.

Are their boards of directors in control of huge corporations? It seems not. Technically directors are elected by the shareholders to oversee corporate management. They are often amply paid but usually do little. Their role is largely honorific, because big corporations want "big names" for their boards. The directors usually attend a few meetings a year, often rubber-stamping management policies. They tend to make decisions based upon incomplete information provided by management. Usually they select top management and personnel, including the CEO.

There is no effective mechanism for oversight of directors. The directors in turn are supposed to oversee management, which they rarely take the time or effort to do. As the massive corporate scandals of the first years of the twenty-first century have demonstrated, corporate executives are often engaged in ploys to increase their own compensation and to ensure themselves "golden parachutes" so that they may accrue substantial rewards even if they fail in their assigned tasks. In companies like Tyco and Adelphia Communications, company executives looted their own companies for personal gain and wealth. Others, as at Enron, played deceptive shell games with company assets with unfortunate results: corporate collapse, loss of jobs, destructive economic ripples through local communities, massive loss of funds by employees and investors—particularly from the portfolios of the elderly and from retirement funds. These corporate robber barons have done more damage than their nineteenth-century forebears.

Certified public accountants are licensed to protect the public interest by monitoring the fiscal activities of publicly held corporations. But the story of Enron and similar corporations shows how the neglect of honest and careful monitoring can lead to disaster. The conflict of interest brought about by accounting firms offering not only accounting but also consulting services to their clients discouraged such firms from providing honest reports of company activities. Accounting/consulting firms did not wish to jeopardize their enormous consulting fees by providing accurate and honest audits. Themselves huge corporations, large accounting/consulting firms such as Arthur Andersen colluded with their large corporate clients in perpetrating fraud and deception, for which Andersen paid the supreme price—corporate death.

Theodore Roosevelt's hope for a separation of business and the state has never been realized. Political leaders continued to receive substantial contributions from corporate giants, and highly paid corporate lobbyists continued to influence local and national politicians. During the 107th Congress, when many of the recent corporate scandals were revealed, either the spouse or child of the Senate Majority Leader, the Senate Minority Leader, the Speaker of the House, and the Majority Whip were active as corporate lobbyists. For corporate giants, large political contributions and the funding of well-connected lobbyists are considered an investment as well as an example of corporate freedom of expression, protected by the First Amendment rights granted to corporations by the Supreme Court. For an "investment" of a few hundred thousand or a few million dollars, politicians can be influenced to ensure legislation that will help various corporations make millions or even billions of dollars—whether or not such legislation is in the public or national interest. As the dissenting opinion in the Supreme Court case of *First National Bank of Boston v. Bellotti* noted, "It has long been recognized . . . that the special status of corporations has placed them in a position to control vast amounts of economic power which may, if not regulated, dominate not only the economy but also the very heart of our democracy, the electoral process."

In corporate America the moral adage "Crime doesn't pay" seems to have little currency. Crime does pay, often handsomely. Punishments are few and often comparatively lenient—considering the crimes and compared with the sentences of offenders whose crimes have hurt no one or very few people. Plea bargains, in which corporations agree to pay fines and awards while admitting no criminal wrongdoing, have grown in number. In May 2002, Warren Buffett told an audience of stockholders in his company, Berkshire Hathaway, that Wall Street loves a crook, that investment bankers have contempt for investors, that stock-option-engorged CEOs are shameless, and that American business is saturated with fraud.

Without effective mechanisms to control corporations, whether by governments, their owners, or other tools of corporate oversight, people are often left defenseless in the face of corporate power and influence. As numerous episodes have revealed, corporate cover-ups, deceptions, and collusions have not only harmed the fiscal well-being of innocent people but their health and lives. A case in point is the Pacific Gas and Electric scandal made famous by the book and later the film *Erin Brockovich*. Not only were carcinogens knowingly dumped in the water supply by Pacific Gas and Electric, but the company for years refused to acknowledge it and disclaimed any responsibility for the high incidence of cancer and other illnesses among local residents, especially children.

Today, like that of the legendary golem, the power of corporate golems seems to have run amok. What we now need are not new laws and regulations or even the enforcement of existing laws. In the long run, such efforts prove ineffective. Rather, we need to reconceptualize the very nature of the corporation. Two quite different traditions can provide guidance in this effort: American law that created the modern corporation, and Jewish law and lore that created the golem legend. From the resources of these two traditions we can develop a response to the question posed above: Can we control the corporations we have created, and if so, how?

According to various Jewish texts about the golem, if one needs to understand why a golem is flawed or has become dangerous, one should investigate how that golem was initially created. In so doing one may discover a means of controlling the golem—and, if necessary, of destroying it. Often the formula for controlling a golem is found in reversing the process that initially brought it into being. For example, according to certain versions of the golem legend, a golem is "activated" by writing certain magical formulas on its forehead, or in writing them on a piece of parchment that is then placed under its tongue. By erasing certain letters from the formula, or by removing the formula from the body of the golem, the golem would thereby be brought under control. Similarly, corporations have been created and animated by certain legal terminology pronounced by the courts. By investigating this terminology we can see where the courts went wrong in the process of creating powerful, often dangerous and destructive corporate golems. By revealing this process, and by reversing some of this language, corporations might be brought under control.

Brandeis and others identified the foundations of corporate abuse in a social and legal shift in the understanding and description of the nature and mission of the commercial corporation. In order to halt corporate malfeasance and restrain potentially dangerous corporate power, he and others recommended returning to earlier understandings and terminology regarding the legal status of the corporation. Initially corporations, like golems, were created to serve the common good. As the Supreme Court stated in the 1906 case of *Hale v. Henkel*, "The corporation is a creature of the state. It is presumed to be incorporated for the benefit of the public." To restrain what Theodore Roosevelt called the "dark power" of corporations, and to restore them as a "benefit to the public," should be the aim of a reconceptualization of their nature.

The judicial process has both the elasticity and the mechanisms necessary for self-correction when unfortunate precedents have been established. Earlier precedents can be reintroduced, new precedents can be established. In the case of corporations, a number of incorrect, unfortunate, and potentially dangerous precedents were adopted. Of these, the most problematic was to grant corporations the protections of certain constitutional rights and privileges previously reserved for natural persons, but without the correlative responsibilities of natural persons. What we now need is the opposite: court rulings or legislation that would remove from corporations the legal rights of natural persons while holding them culpable and accountable for their misdeeds and for the damages they inflict upon people and property.

The *Santa Clara* case, by which corporations were afforded the status, rights, and legal protections of natural persons under the Fourteenth Amendment, established a bogus precedent. The court never actually made such a ruling in this case. Although some Supreme Court justices accepted the alleged position of the Court in *Santa Clara* as precedential, others did not.

In a variety of cases, various Supreme Court justices asserted that the ruling in *Santa Clara* mistakenly confused corporations with natural persons. In the 1977 case of *First National Bank of Boston v. Bellotti*, for example, Justice Rehnquist noted that "The Court decided at an early date [in 1886 in Santa Clara], with neither argument nor discussion, that a business corporation is 'a person' entitled to the protection of the Equal Protection Clause of the Fourteenth Amendment. . . . Nevertheless, we concluded soon thereafter that the liberty protected by that Amendment 'is the liberty of natural, not artificial persons.'" Rehnquist further noted that the Court's denial of Fifth Amendment rights to corporations (in the 1944 case of *U.S. v. White*) demonstrated that "the mere creation of a corporation does not invest it with all the liberties enjoyed by natural persons."

Similarly, Justice Black in 1938 said bluntly, "I do not believe the word 'person' in the Fourteenth Amendment includes corporations. . . . A constitutional interpretation that is wrong should not stand. I believe this Court should now overrule previous decisions which interpreted the Fourteenth Amendment to include corporations. Neither the history nor the language of the Fourteenth Amendment justifies the belief that corporations are included within its protection. . . . The records of the time can be searched in vain for evidence that this amendment was adopted for the benefit of corporations. . . . The history of the amendment proves that the people were told that its purpose was to protect weak and helpless human beings and were not told that it was intended to remove corporations in any fashion from the control of state governments. . . . The amendment was intended to protect the life, liberty, and property of human beings. The language of the amendment itself does not support the theory that it was passed for the benefit of corporations. . . . No word in all this amendment gave any hint that its adoption would deprive the states of their long-recognized power to regulate corporations."

Black's recommendation that the Court overrule and reverse the precedent allegedly established in *Santa Clara*, and Rehnquist's claim that subsequent courts never really accepted that precedent, might be extended to suggest that the Court completely divest corporations of all constitutional rights and privileges granted them by the courts. These rights were clearly meant to protect natural persons, human beings, and not artificial persons—that is, corporations. Stripping corporations of the rights of natural persons would severely limit their power.

The depersonalization of the corporation would help restore the rights of natural persons that have been trumped by the granting of constitutional protections to corporate entities. Indeed, political activists already have begun to introduce "Ordinances to Deny Corporate Personhood" in local communities throughout the United States. This and related endeavors may grow to the point of persuading the Supreme Court to rescind earlier precedents granting the rights of natural persons to corporations and to build upon those precedents that refused to do so.

A further critical step toward asserting government control over corporate activities is to limit the influence of corporations on government. Without the protections of certain First Amendment rights, corporations would no longer be able to compromise the democratic electoral process or to have undue influence on government policy through political donations to candidates, political parties, or legislators. As the dissenting justices in *First National Bank of Boston v. Bellotti* wrote, "The State need not permit its own creation to consume it."

Relieving corporations of the rights and privileges of natural persons, and limiting their influence over the government, would be important steps in placing corporations under effective oversight. Yet these measures alone might not be enough to control corporate power and malfeasance. Corporations should be stripped of the legal status of persons, both natural and artificial.

In retrospect it has become increasingly clear that British common law erred by granting to corporations the status of "artificial persons." Although common law carefully distinguished between artificial and natural persons, it could not have foreseen a time when such a distinction would be eroded. Having granted personhood to corporations, early British common law could not have imagined how its application to international conglomerates could help them wield more economic power than many nation-states. In Blackstone's time most corporations were small communal, ecclesiastical, and fraternal organizations, aimed at serving the public good. Blackstone undoubtedly would have been horrified at the violations of the public trust characteristic of the activities of many of today's enormous commercial corporations.

Earlier I referred to a decision in Jewish case law by the seventeenth-century rabbi Zevi Ashkenazi. Ashkenazi rejected the claim that a golem might be counted in a quorum for prayer. This decision may be taken as a precedent in the question of whether an artificial creature can be granted the legal status of a person. Ashkenazi ruled that it could not. From the perspective of Jewish law, considering a corporation or a golem as a person creates an unnecessary legal fiction. In this view a corporation is not a person, in form, nature, or essence. Consequently, there is no reason to grant it the legal status of a person. Had British common law taken a similar stance regarding corporations and denied them the legal status of a person, many of the abuses of corporate power with which we now contend might have been substantially reduced.

Anglo-American law should rescind the status of large corporations as artificial persons and consider them merely as golems, as artificial entities, and not as legal persons in any sense of the term. In this regard the remarks of Delphin Delmas, the attorney for Santa Clara County in the *Santa Clara* case, are prescient. Arguing before the Supreme Court, Delmas said, "The shield behind which [the Railroad] attacks the Constitution and the laws of California is the Fourteenth Amendment. It argues

that the amendment guarantees to every person within the jurisdiction of the state the equal protection of the laws; that a corporation is a person; that therefore, it must receive the same
protection as that accorded to all other persons in like circumstances. . . . The whole history of the Fourteenth Amendment
demonstrates beyond dispute that its whole scope and object was
to establish equality between men . . . and not to establish equality between natural and artificial beings. . . . [The] mission [of
the Fourteenth Amendment] was to raise the humble and the
downtrodden and the oppressed to the level of the most exalted
upon the broad plain of humanity—to make man equal of man;
but not to make a creature of the state—the bodiless, soulless,
and mystic creature called a corporation—the equal of the creature of God."

The dominant position in Jewish law is to treat corporations
not as persons but as partnerships. As such, corporations would
enjoy none of the rights of persons except property ownership.
There would be no "corporate shield" for managers and directors
to hide behind. There would be no intermediary entity to evade
individual moral and legal responsibility for corporate misdeeds.
Like a golem, when a corporation has outlived its purpose, it
could be destroyed. When a corporation poses too great a danger
to life or property, it could be dismantled.

Jewish law views corporations as individuals acting in a
group, in a partnership. This aggregate idea of corporate partnership is well known in corporate theory. In this view a corporation is a group of individuals who engage in a particular type of
contractual relationship with one another. There is only an association, no separate corporate "person." Managers and directors
are "partners"; stockholders, however, are not. To consider
stockholders as partners would expose them to liabilities they

might not be willing to risk, especially if they have no control over the management of corporate activities. Indeed, to impose personal liabilities on stockholders—which they currently do not have beyond the cost of their stock—would cripple the ability of even the best corporations to do business. According to Jewish law, stockholders are not treated as partners, owners, or managers (unless they happen also to be such) but as creditors who put their money at the disposal of corporate management. If the money is lost, it need not be repaid. If a profit is made, they share in the benefits.

Perhaps because he knew what had happened to his ancestor, Rabbi Elijah of Helm, Zevi Ashkenazi had no qualms about dismantling a golem when it threatened its creators or society at large, or even when it outlived its mission. Perhaps Ashkenazi would have expressed a similar view of corporate golems. American law too has precedents for dealing with corporate golems that have caused or threaten to cause damage to life or property; or even corporations that have outlived their purpose or fulfilled their mission.

A case in point was decided in New York in 1890, when the highest court in the state revoked the corporate charter of a company that had "violated its charter and failed in the performance of its corporate duties." In *People v. North River Sugar Refining Corporation*, the court declared unanimously: "The judgment sought against the defendant is one of corporate death. The state which created, asks us to destroy, and the penalty involved represents the extreme rigor of the law. . . . Corporations may, and often do, exceed their authority . . . but when the transgression has a wider scope, and threatens the welfare of the people, they may summon the offender to answer for the abuse of its franchise and the violation of its corporate duty."

As is clear from this case, the state that creates corporate golems has the power to destroy them, and it can exercise that power when justified. Corporations therefore need not have the "legal immortality" that Blackstone referred to. The ultimate punishment for corporate malfeasance is corporate execution by the state, corporate death. In this way the state can exercise ultimate control over corporate entities. In this way corporate golems may be deactivated and controlled before they become too dangerous.

That corporations may be "executed" for their misdeeds implies that they can be held culpable for their crimes. This approach rejects the opinion that considers corporations to be amoral. Milton Friedman, a Nobel Prize Laureate in Economics, and others have argued that the social responsibility of a business corporation consists primarily in its ability to serve its shareholders by producing profit for them. In this view corporations are amoral; they have "no pants to kick, no soul to damn." Others maintain that as an "artificial person," a corporation cannot be considered a moral agent and therefore cannot be held responsible for the intention to commit a crime or for the crime itself. But in 1909, in the Supreme Court case of *New York Central and Hudson River Railroad Company v. United States*, the issue of corporate responsibility came to a head. The questions before the Court were: Can a corporation commit a crime? Can a corporation have criminal intent? How can a corporation be punished for having committed a crime?

Until this case the dominant view, as stated in *Blackstone's Commentaries*, was that "A corporation cannot commit treason, felony, or other crime in its corporate capacity, though its members may, in their distinct individual capacities." In this case, however, the Supreme Court invoked a more "modern" approach. It held that corporations might indeed commit crimes and be punished for so doing. In its decision the Court noted, "If, for example, the invisible, intangible essence or air which we term a corporation can level mountains, fill up valleys, lay down iron tracks, and run railroad cars on them, it can intend to do it, and can act therein as well viciously or virtuously." Articulating the view that corporations can not only act but can also have intent, Chief Justice Field said, "We think that a corporation may be liable criminally for certain offenses of which a specific intent may be a necessary element. There is no more difficulty in imputing to a corporation a specific intent in criminal proceedings than in civil. A corporation cannot be arrested and imprisoned in either civil or criminal proceedings, but its property may be taken, either as compensation for a private wrong or as punishment for a public wrong."

According to the Court, corporations themselves could be culpable for deeds performed by their agents. Corporate agents might be indicted, tried, and punished for their own misdeeds, but corporate entities were also liable for their intentions and actions when executed by their authorized agents. Why? "If it were not so, many offenses might go unpunished and acts committed in violation of law." Finally the Court observed, "While the law should have regard to the rights of all, and to those of corporations no less than to those of individuals, it cannot shut its eyes to the fact that the great majority of business transactions in modern times are conducted through these bodies, and particularly that interstate commerce is almost entirely in their hands, and to give them immunity from all punishment because of the old and exploded doctrine that a corporation cannot commit a crime would virtually take away the only means of effectually controlling the subject-matter and correcting the abuses aimed at." The ruling in this case, however, never became dominant law.

Like modern golems, modern corporations can and must be brought under control, lest they harm their creators and society at large. By "depersonalizing" business corporations, by bringing them under the control of appropriate laws and regulations, by holding them and their agents culpable and liable for their misdeeds, corporate malfeasance can be restrained. Thus corporations might be deterred from evolving into the Frankensteinian monsters that Justice Brandeis wrote about. The implementation of this model would represent a complete reversal of the model desired by the railroad companies at the end of the nineteenth century. They wanted corporations to have the rights but not the liabilities of natural persons. In this new model, however, corporations would have none of the rights or protections of law afforded to natural persons (except property rights), but would nonetheless be subject to the same legal culpabilities and liabilities of natural persons.

Like his ancestor, Judah Loew of Prague, Justice Brandeis understood that artificial beings had to be controlled, and that they must be destroyed once their power has grown too great to restrain. Like golemic death, corporate death is sometimes the only means of treating entities that have eluded our grasp and taken on a life of their own. When the entities we have created to help and defend us run amok, when they threaten to deprive us of our liberties, our health, and even our lives, it is time either to control them or return them to the elements from which they came.

Chapter Ten

The Honey
and the Sting:
The Golem Meets
Frankenstein

In March 1980, four days after the Supreme Court affirmed Chakrabarty's right to a patent on the new genetically engineered bacterium he had created, a letter from leaders of the American Protestant, Catholic, and Jewish communities was sent to President Jimmy Carter. They asked the president to consider the many ethical and public policy issues related to genetic engineering that had been neglected by the Court in its decision. In response, President Carter instructed the President's Commission for the Study of Ethical Problems in Medicine and Biomedical and Behavioral Research to undertake a study of "the social and ethical issues of genetic engineering with human beings." In November 1982 the commission issued its findings in a report titled *Splicing Life*.

The report observed: "Like the tale of the Sorcerer's apprentice or the myth of the Golem created by lifeless dust by the 16th-century rabbi, Loew of Prague, the story of Dr. Frankenstein's monster serves as a reminder of the difficulty of restoring order if a creation intended to be helpful proves harmful instead. Indeed, each of these tales conveys a painful irony: in seeking to extend their control over the world, people may lessen it. The artifices they create to do their bidding may rebound destructively against them—the slave may become the master."

Concerns such as those raised by the commission have been articulated in recent decades by writers, artists, theologians, ethicists, scientists, and others who have dealt with the implications of developments in bioengineering, reproductive biotechnology, robotics, and related areas. Like the members of the commission, they have often associated the legend of the golem with the story of Frankenstein because both are concerned with the creation of artificial life and its implications.

The legend of the golem and the story of Frankenstein are among the most influential and powerful epoch myths of modern times. Each is set in a particular time and place. Yet, like all enduring myths, they speak to the human condition at many times and places, including our own. Both are stories that address universal problems: the mystery of life, the nature of human creativity, the danger of power, the relationship between human beings and nature, and the relationship between human beings and the artifacts they have introduced into the world.

It has been said that some stories are true though they never happened, and other stories are not true though they actually have occurred. Although Rabbi Loew was an historical figure, he never created a golem. The story of Rabbi Loew and the golem of Prague is a work of fiction. In the case of *Frankenstein*, both the creator and the creature are fictional characters. Mary Shelley created Victor Frankenstein, who in turn created a nameless being in the pages of her 1818 novel. Nonetheless, both the legend of the golem of Prague and the story of Frankenstein are true because the issues they raise are perennially valid and relevant, and because creatures that bear their pedigree are real and now dwell among us.

The golem legend and the story of Frankenstein are often treated as if they were really one and the same story. But a careful examination indicates that they are very different; each offers a radically different response to many of the questions they raise. As we have seen, elements of the Frankenstein story have been incorporated into recent versions of the golem legend, thereby distorting its nature, meaning, and message. This is especially the case in contemporary literature and works of popular culture. In such depictions of the golem—for example, the Galactic Golem of *Superman* comics, and the "Golem robot [that] terrorizes Gotham City" in *Batman Beyond* comic books— the Frankenstein-like golem is portrayed as an out-of-control, malevolent monster. The golemic monster found in these contemporary works is closer to a facsimile of Frankenstein than to the golem of Prague or its predecessors in classical Jewish legend. Precisely because the stories of the golem and of *Frankenstein* have been portayed as being the same, it is important to demonstrate how they differ. The first issue to examine is the contrast between Judah Loew and Victor Frankenstein.

Unlike Victor Frankenstein, Judah Loew of Prague is a person of profound religious faith and moral virtue, a communal leader and a seasoned scholar. Aged and mature, he creates the golem with hesitancy but out of a perceived urgent necessity. In creating his golem, Loew follows hallowed traditions, using powers acquired through decades of study and a life of holiness. His aim is to benefit his family and to protect his people. He creates by invoking a sacred legacy that allows him to fulfill a religious mandate to "imitate the ways of God." In penetrating the mysteries by means of which God created life, Judah Loew is brought into cohesion with the divine. He serves, as the Talmud puts it, as "God's partner in the work of creation."

Bearing responsibility for his creature, Loew never lets it elude his control for too long. He feeds, dresses, and cares for his golem, teaching it the ways of the world. Loew acts under the aegis of prescriptions drawn from a sacrosanct ancient faith and tradition. Astonished by and in awe of what he has accomplished, Loew exhibits neither hubris nor arrogance. When his creature has completed its mission, Loew sadly lays it to rest. Just as Loew serves his Creator, his golem remains obedient to and under the control of its creator, Rabbi Loew.

Unlike Judah Loew, Victor Frankenstein is a self-obsessed "pale student of unhallowed arts." Barely a novice in his pursuits, he is scarcely more than a coddled adolescent. A recluse, untutored in the ways of the world, Frankenstein is driven by pride to demonstrate his power. Although he is intellectually precocious, his acquired knowledge surpasses his wisdom. Emotionally infantile, he gives little thought to the consequences of his actions. Both his deeds and the creature he brings into being are described as hideous, evil, deformed. Frankenstein is repulsed both by what he has done and by the creature he has brought into the world. Frankenstein never cares for his creature, leaving it to its own devices. Only after the creature has unleashed its destructive fury against the innocent people whom Frankenstein most loved does he take action—more for revenge than out of moral responsibility or a desire for justice. Frankenstein cannot control the creature he has brought into the world. It eludes his grasp. He falls under its control, the servant having become the master.

The legend of the golem and the story of Frankenstein are far from being the same story. They are radically different renditions of a common theme. In each of these stories, the motivation for creating artificial life is different, as is the method, the goals, the character of the creator, the nature of the being created, the ability of the creator to control his creature, and the creator's assumption of responsibility for the actions of his creature. The stories differ because each is informed by a distinctive cultural and intellectual tradition. The golem legend flows from Jewish religious and cultural tradition. The story of Frankenstein draws upon motifs embedded in ancient Greek mythology and nineteenth-century Romanticism.

The subtitle of *Frankenstein* is "The Modern Prometheus." In Greek mythology, Prometheus is a demigod worshiped by craftsmen. (The Greek word for craft, *techne*, is the origin of the English word *technology*. The first cloned horse was named Promethia.) In his book *The Human Use of Human Beings* (1950), Norbert Wiener described Prometheus as "the prototype of the scientist." Like Wiener, Mary Shelley understood the modern scientist, epitomized by Victor Frankenstein, as a modern Prometheus.

In Greek mythology, Prometheus steals fire from the gods and gives it to humankind without divine sanction. For this act, Zeus punishes him by chaining him to a rock where an eagle feeds daily on his liver. Like Prometheus, Frankenstein is described as invading the divine realm and appropriating a power he has no right to possess. Frankenstein presumes to have divine powers—the ability to create life—and, like Prometheus, he is punished accordingly. Like Prometheus, Victor Frankenstein enjoys a victory (hence the name Victor) over the divine, for which he pays an enormous price. By rebelling against the gods instead of seeking communion with the divine (like Judah Loew), Frankenstein—like Prometheus—unleashes a perilous force upon the world, which results in his own demise.

Shelley was undoubtedly influenced by the Roman as well as the Greek versions of the myth of Prometheus. The Roman version is found in Ovid's *Metamorphoses*. There (as in the Hebrew Bible and later in the golem legend) human beings are created and manipulated into life from earth. In the Bible it is God who creates humankind. In rabbinic tradition the creation of the golem is believed to have been sanctioned by a divine mandate. In contrast, Ovid describes how Prometheus, the Titan, without permission from the gods, takes "earth [that] was metamorphosed into man."

In her description of the modern Prometheus, Shelley rejects Lord Byron's reading of the myth of Prometheus in his poem "Prometheus Unbound." Byron saw the legend as having a salvific message. For Shelley, however, Victor Frankenstein, the modern Prometheus, is not a savior. His efforts are not for the ultimate benefit of humankind. Rather, Frankenstein's actions are an expression of the sin of hubris, for which he deserves and receives the inevitable punishment.

Throughout *Frankenstein*, the ancient Greek preoccupation with the moral vice of hubris and its inevitable catastrophic implications are reiterated. Because of his hubris, Victor Frankenstein's life becomes a chain of tragedies. In *Frankenstein*, as in

197

the ancient Greek dramatic tragedies, catastrophes inevitably follow the protagonist's pride-driven attempt to alter destiny, to modify the order of nature, to usurp something of the divine. Motivated by hubris, Frankenstein declares, "I will pioneer a new way, explore unknown powers and unfold to the world the deepest mysteries of creation." Unlike Judah Loew, who acts as "God's partner," Frankenstein, like Prometheus, seeks to storm the heavens to usurp power from the gods. Unlike Loew, who utilizes the sacred arts, Frankenstein employs the "unhallowed arts."

Reflecting upon his deeds, Frankenstein finds them and the being he has created evil and odious. "Frightful must it be," he says, "for supremely frightful would be the effect of any human endeavor to mock the stupendous mechanism of the creator of the world. His success would terrify the artist; he would rush away from his odious handiwork, horror-stricken. He would hope that, left to itself, the slight spark of life which he had communicated would fade; that this thing, which had received such imperfect animation would fade; that this thing . . . would subside into dead matter."

For Frankenstein and the monster he had created, there can be only a tragic end. In a relationship dominated by mutual enmity, loathing, and disgust, the only possible resolution is a contest to determine who can best torment and eventually destroy the other. As the creature tells Dr. Frankenstein, "You my creator, detest and spurn me, thy creature, to whom thou art bound by ties only dissoluble by the annihilation of one of us."

The story of Frankenstein has become a metaphor for the profound anxieties that people display as they encounter rapidly unfolding developments in contemporary science and technology as well as the phenomenal growth of corporate power. In his book *Artificial Life* (1992), Steven Levy writes, "Those steering artificial life toward the creation of autonomous, evolving organisms, truly will become successors to the fictional Victor Frankenstein, who was destroyed not so much by his own creation as by his willingness to tamper with the justifiably forbidden." As Levy notes, early in the development of "artificial life," physicist James Doyne Farmer warned of the "bugaboo of Frankenstein"—that the scientist, intoxicated by the hubris connected with the ability to create life, would proceed unencumbered, assuming no responsibility for the dangerous and potentially catastrophic outcomes of an endeavor that could destroy not only himself but many innocent people as well.

As we confront our own future, the figure of Victor Frankenstein and the monster he created looms large. Many people today see our present and future world through the lens of the Frankenstein story. They affirm the lessons of *Frankenstein*: that humans should not create artificial beings, not "play God," refrain from tinkering in the toolbox of nature, and avoid doing what is unnatural. *Frankenstein* admonishes us to suppress our Promethean urges lest we unleash an inevitable tidal wave of tragedy and catastrophe upon ourselves and our society. In contrast, the golem legend encourages us to utilize sanctified power to protect and enhance our own lives, to imitate the ways of God, to fulfill the mission of improving the world bequeathed to us, to temper power with moral vision, and to maintain control over our creations.

In dealing with the challenges, opportunities, and dangers posed by contemporary science and technology, two models present themselves: Judah Loew and Victor Frankenstein, the golem of Prague and Dr. Frankenstein's monster. Which we choose to emulate may not only determine what kind of future we have but whether we have a future at all. The golem legend offers us a poignant alternative both to the story of Frankenstein and to the outlook of those who anticipate the future with ebullient optimism, as we confront the spiritual and moral challenges of our biotech century. The ancient and medieval wisdom offered by the legend of the golem and by the Jewish tradition that spawned it can serve as a prudent pilot in helping us navigate a safe journey through the minefield of social and biological engineering in which we now stand, where a step one way may reveal a panacea while a step in the other may precipitate a calamity.

The golem legend was created and nurtured by the rich and venerable Jewish spiritual and literary tradition that begins with Hebrew Scripture's account of the creation of the universe, of our world, and of humankind. Not meant to be a scientific account of the origin of the universe or of life on earth, the biblical story of creation aims to assert certain attitudes about human nature and about the relationship between human beings and nature. When seen within its historical context, the biblical story of creation represents a radical departure from and a revolutionary polemic against the attitudes and views of ancient Near Eastern culture and religion. Although biblical attitudes toward nature reflect the clash of ancient cultures, they nonetheless address contemporary attitudes and issues as well.

The religion of biblical Israel emerged in the ancient Near East against the backdrop of the great cultures of Egypt and Babylonia. Ancient Babylonian religion claimed that nature is divine, that the elements of nature—such as the sun, the moon, and the sea—are gods. Consequently, since nature is divine, it ought to be worshiped. In contrast, biblical religion teaches that nature is not divine but a creation of God. Therefore nature is not to be worshiped. Rather than seeing human beings as subservient to nature, the Bible considers them as stewards of nature. Because humans are the only creatures described by Scripture as having been created in "the image of God," they are given dominion over "the whole earth" (Genesis 1:26).

Ancient Near Eastern religions asserted that not only human beings but even the gods are subject to destiny, that everything is predetermined, that all events ever to occur have been inscribed for eternity in the Tablets of Destiny. From this perspective, neither the gods nor human beings are free moral agents. In contrast, biblical religion claims that God brought the world into being in an act of creative volition. Rather than being a force of nature, the Bible describes God as transcending nature. And Scripture describes human beings as the highest created beings. Precisely because human beings have been created in the divine image, they, like God, are free moral and creative agents. Although they are a part of nature ("Dust you are and to dust you will return"—Genesis 3:19), human beings are also apart from nature. Neither servants of nature nor puppets of destiny, human beings, like God, transcend nature. They are intelligent and creative beings who can alter the world that God created.

The claim that our lives and our actions are determined by destiny and fate reverberates throughout ancient Babylonian religious teachings. It later shaped the message of the Greek tragedies that informed *Frankenstein*. It resurfaced in the medieval idea of "Fatima" or fate, and in the practice of the medieval "science" of astrology. In our own time, fatalism has reappeared in the notion that our genetic predispositions determine who we are and who we will become. As James Watson, the co-discoverer of DNA, put it, "We used to think our fate was in the stars. Now we know in large measure our fate is in our genes."

Fatalism also informs the views of scientists like Ray Kurzweil and Hans Moravec, who consider the transition of human life to machine-life as both inevitable and desirable. Scientists such as Gregory Stock write about "redesigning humans" in "our inevitable genetic future." "Behavioral genetics" is already being used to explain as well as to justify all kinds of human behavior.

In stark contrast to the fatalistic, deterministic approach to human behavior, Judaism from its beginnings has taught that freedom of moral choice is an endemic human characteristic, that our deeds are not determined, that without moral choice there can be no moral responsibility. Maimonides, for example, taught that though we have certain behavioral dispositions, they can be controlled through discipline and through the cultivation of moral virtue. Judah Loew taught that when Scripture describes human beings as having been created in "the image of God," it refers to moral and creative volition. From this perspective, if we cannot control our own dangerous inclinations, how can we expect to control the artifacts we have created?

Contrary to popular modern opinion, the biblical view of the human relationship with nature did not stifle scientific and technological development but made them both possible and desirable. Once nature was rejected as an object of worship, once nature was desacralized, and once human transcendence, freedom, and creativity were affirmed, the human prerogative to investigate and modify nature for human purposes became sanctioned and encouraged. Had nature remained a divine force worthy of worship, rather than a work-in-progress requiring human development, technology would have been condemned.

In the biblical view, as interpreted by the Talmudic rabbis, God created the world but left it unfinished. In rabbinic parlance, human beings are mandated to serve as "God's partners in the work of creation." The world has been "created to be made"—and the human mission is to work toward completing the process of creation begun by God. By so doing, human beings are neither impersonating God nor "playing God." Rather, they are articulating their having been created in the image of God. Beneficial human interventions in nature fulfill the divine mandate to human beings to subdue nature and to establish their dominion over it (Genesis 1:28).

When Prometheus breaches the domain of the gods and steals fire from them, he is harshly punished. In contrast, the Talmud describes God as providing the means for Adam to "invent" fire. According to a rabbinic legend, God had compassion upon Adam when God saw him frightened and shivering in the night. So God bestowed upon Adam elements of God's own knowledge and creativity. This inspired Adam to "invent" fire by rubbing two stones together. Later, the legend continues, Adam crossbred a donkey and a horse to create a mule, a new species of animal—the first product of interspecies bioengineering.

Commenting on the creation of animal life-forms by human beings (specifically the Talmudic account of the creation of a calf by two rabbis), the thirteenth-century Talmudic commentator Menahem ha-Meiri writes, "Even if one knew how to create creatures without natural procreation, as is known in the books of nature, he may engage in such activity, since anything natural is permitted and not forbidden."

Similarly, Rava's creation of a golem—an anthropoid—as recorded in the Talmud, is never condemned by the ancient or medieval rabbis. The propriety of the human creation of new or existing life-forms through "artificial" means is sanctioned. It is not considered to be an illicit dabbling in the "unhallowed arts." In creating life from dust, a living being from inert matter, Rava imitates the way of God.

In the writings of Judah Loew we find a strong endorsement of the human ability to improve upon nature, to create new creatures, and to alter nature for human purposes: "Everything that God created requires [human] repair and completion." For Loew, such activities are not unnatural but rather are natural ways of extending nature, of helping nature realize its dormant potential. For Loew, human creativity is a natural way of transcending nature. In *A Winter's Tale*, Shakespeare wrote:

So over that art
Which you say adds to Nature
Is an art That Nature makes . . .
This is an art
Which does not mend Nature,
Change it rather: but
The art itself is Nature.

While Jewish tradition sanctions and even encourages human intervention in nature, it recognizes that the license to alter nature comes with certain limits. Without boundaries and guidelines, risk may easily escalate to danger and catastrophe. Human ingenuity may readily degenerate into arrogance and pride, for as

Scripture reminds us (and *Frankenstein* demonstrates), "Pride comes before a fall." The human creator may be corrupted by power. The creature may assume a life of its own, reel out of control, and wreak irreparable havoc and destruction. Jewish tradition recognizes that creativity and innovation entail risk, but they need not portend catastrophe.

The Talmudic rabbis remind us that from the bee we can receive either its honey or its sting. The bee's honey is beneficial and sweet, but its sting can be harmful, even lethal. Should we forgo the sweetness of the honey because of the danger of its sting? *Frankenstein* would say yes, because tampering with nature brings inevitable catastrophe. The golem legend says no, because the actual and potential benefits may be too immense to be readily surrendered, and by exercising prudence the dangers can be mitigated.

In our time the genie has already been released from the bottle, and the challenge we now face is not whether to free it, but how to control it, and whether additional genies should be released or confined to their bottles. In a similar vein, the British poet W. H. Auden wrote:

> The sense of danger must not disappear:
> The way is certainly both short and steep
> However gradual it looks from here;
> Look if you like, but you will have to leap.

To help alleviate the dangers, Jewish tradition offers certain recommendations for the application of technological knowledge and know-how. One such suggestion is that knowledge be nourished by wisdom.

Science comes from the Latin word meaning knowledge; *technology*, as noted earlier, comes from the Greek word meaning craft. Science is what we know; technology is what we can *do* with what we know. Creation of a golem requires both knowledge and technical ability—but it also involves wisdom. Wisdom is knowing what to do with what we know and with what we can do. Wisdom is also knowing when, whether, and why to do what we can do. About wisdom, Jonas Salk said: "Wisdom is the capacity to make judgments that, looking back upon them, seem to have been wise."

Science and technology have granted us the knowledge and the technical ability to do many wondrous things, but they have not taught us wisdom. While ancient and medieval science dealt with wisdom as well as knowledge, modern science and technology have largely turned their backs on questions of "why," on issues of human meaning and morality, focusing instead on questions of "how to." We are therefore compelled to ask the question posed by T. S. Eliot, "Where is the wisdom we have lost in knowledge?"

Wisdom asks us to embrace judiciousness over opportunism, to question answers as well as to answer questions, to survey the panorama rather than to rest content with tunnel vision, to consider long-term consequences rather than to focus only on short-term needs. As the Talmud puts it, "One who has acquired wisdom has acquired everything. A person who has acquired wisdom, what does he lack? A person who lacks wisdom, what has he acquired?"

In Hebrew, each letter of the alphabet is also a number. According to one spelling of the Hebrew word golem, its numerical equivalent equals 73 (g-l-m = 3 + 30 + 40 = 73). Perhaps not by chance, the Hebrew word for wisdom—*hokhmah*—also equals 73. The message here is that the creation of a golem requires wisdom as well as knowledge and technical skill. Precisely because the golem lacks wisdom, its creator must possess it. In our biotech century, the fundamental question is not, Should we do what technology enables us to do? Rather it is, Is it wise to do what we can do?

The Talmudic treatise *The Ethics of the Fathers* uses the term *golem* to denote a person who is the opposite of one who is wise. Here *golem* refers not to a humanly created artifact but to a person who lacks wisdom. According to Maimonides' interpretation of this text, a golem is a person who has knowledge and good intentions but who is morally and intellectually confused. As Mary Shelley describes Victor Frankenstein, he is such a person. For Maimonides, a wise person is one with clarity of knowledge and purpose as well as a highly developed moral and spiritual character. Such a person was Judah Loew of Prague.

Centuries ago Plato taught that "wisdom is the essence of virtue." The quest for wisdom entails the cultivation of moral virtue. Our technological achievements must be informed by our ethical values and concerns. Victor Frankenstein's tragedy was a product of his intellectual and moral confusion, of underdeveloped technological ability being applied without moral virtue.

Like Victor Frankenstein, the modern Promethean scientist has proceeded with scientific and technological developments without moral restraint or responsibility. Tom Lehrer satirized this attitude in his lyrics about the German rocket scientist Wernher von Braun:

> "The rockets go up
> I don't care where they come down
> That's not my department,"
> Says Wernher von Braun.

So it is noteworthy that the U.S. government project to sequence the human genome was the first large, government-funded scientific project to include a budget-line for dealing with the ethical implications of the scientific work being undertaken. Realizing the inevitable ethical implications of this project, moral issues were integrated into the unfolding of scientific developments.

According to the Bible, the first important technological enterprise undertaken by human beings was the construction of the Tower of Babel (Genesis 11). While Scripture informs us that those who built the tower were punished for their sins, the text does not disclose the nature of their moral flaws. For this reason the Talmudic rabbis and medieval Jewish commentators offer a number of suggestions as to what those sins might have been. Their observations speak to the question of tempering technological ability with wisdom.

According to rabbinic legend, when the tower was being built and was already quite high, it was difficult to get bricks to the top of the construction site in order to build still higher. At that point, if a brick fell and broke as it was being hoisted to the top, all the workers cried and mourned its loss, saying, "How shall we get a brick to replace it?" But when one of the workers fell off the tower and died, no one paid attention because workers were plentiful and easy to replace. The message here is clear. The technocratic mentality, epitomized by a combination of technological hubris and pecuniary greed, can readily lead to the commodification of human beings. The sin of those who built the tower was not only their devaluing of human life but their esteeming the artifacts and commodities they had created over individual human life.

The moral message elicited by the Talmudic rabbis from the biblical story of the Tower of Babel is one that rings true today.

Technological achievement that is not guided by a moral compass is ultimately self-defeating and self-destructive. To devalue human life in an effort to enhance it is a pyrrhic victory, the kind epitomized by the exploits of Victor Frankenstein. It is not the way of Judah Loew of Prague.

The biblical tale of the Tower of Babel is meant to satirize those who built the tower. In ancient Babylonian, *bab-el* meant "the gate of god." The biblical text tells us that the word *bab-el* is related to the Hebrew word *balal*, meaning confusion. This seems to be the origin of the English verb *to babble*. For the biblical author, the builders of the tower thought themselves to be gods by virtue of their technological achievement; in fact, like Frankenstein, they were only deluded and confused by their own pretentious pride. They had adequate technological skill but inadequate moral wisdom.

A rabbinic legend tells that Nimrod, the king of the Babylonians, directed the building of the tower. The rabbis related his name to the Hebrew word *mered*, meaning rebellion, and suggested that his goal of building the tower was to rebel against God. Like Prometheus, Nimrod wished to invade the realm of the divine and usurp divine power. This legend describes Nimrod as trying to murder God to establish his own sovereignty over the earth. It tells how he ordered his archers to stand on top of the tower and shoot arrows into heaven in order to kill God, and how God sent them back dripping with blood to make Nimrod think he had accomplished his goal. In this view, technological achievement runs the spiritual risk of human beings believing they have murdered God and have replaced him as ruler and creator of the world. In other words, for the Talmudic rabbis, human technological achievements may confuse people into believing they have become divine. The spiritual danger of technological achievement is not simply that human beings may "play God" but that they may deceive themselves into behaving as if they were omnipotent deities who can control nature as well as human beings for their own purposes. Although the rabbis sanctioned the imitation of God, they drew the line at human beings impersonating God.

Among the ways of imitating God is knowing when to stop. Wisdom involves knowing when to create, when not to create, and when to stop creating. As a surgeon once told me, "I don't get paid according to how much I cut. I get paid for knowing when to stop." Wisdom means recognizing our physical, moral, and spiritual limitations. As the poet Richard Wilbur observes, "Limitation makes for power; the strength of the genie comes from his being confined in the bottle."

According to the biblical story, after the six days of creation God stops and rests. In other words, the challenge is not only to know how and when to create, but also when to stop creating. Commemorating the Sabbath on the seventh day of creation is a celebration not only of God's act of creating the universe but also of God's wisdom in knowing where and when to stop creating. The legend of the golem and the Jewish tradition that spawned it teach that human creativity is sanctioned and encouraged, but that we must also have the moral wisdom to know when to stop creating.

When the entities we create to help and defend us threaten to harm or destroy us; when the artifacts we introduce into our world run amok; when the creatures we invent to serve us end up as our masters; when power and greed diminish our moral clarity; when the lust for knowledge clouds our quest for wisdom; and when we start to value our products more than ourselves— it is time to stop creating. When we choose Frankenstein over Judah Loew, when we practice the "unhallowed arts," it is time to stop creating.

Isaac Asimov's "Three Laws of Robotics" might be applied not only to artificial beings such as androids and robots but also to artificial beings like corporations and to the biotech industry as a whole:

1. A robot [or any other artificial being, including a corporation or a bioengineered entity or process] may not injure a human being, or through inaction allow a human being to come to harm.

2. A robot [or any other artificial being] must obey the orders given it by human beings except where such orders would conflict with the First Law.

3. A robot [or any other artificial being] must protect its own existence as long as such protection does not conflict with the First or Second Law.

Where do we go from here? Consider a Hasidic tale: A certain king heard that a wise man dwelt in his kingdom. Wanting to test his wisdom, the king had the wise man brought before him. The king held out his hand and told the wise man that he had a bird hidden in his fist. It was the wise man's task to determine whether the bird was alive or dead. The wise man knew that if he said the bird was alive, the king would crush it to death. But he also realized that if he said the bird was dead, the king would open his hand and let the bird fly free. The wise man knew as well that if he made the wrong choice, the king would have him executed. What did the wise man say? He thought for a moment and replied, "Your majesty, the answer is in your hands."

Our future is in our hands as our biotech century unfolds. Whether our age will be the age of Dr. Frankenstein and his monster or the age of Judah Loew and his golem is up to us. As he protected the Jews of Prague long ago from their enemies, may the golem protect us from ourselves.

A Note on Sources and a Personal Note

The historical development of the golem legend and its contemporary implications have occupied my thinking, teaching, research, and writing for many years. During the time I have dealt with the legend and its implications, new developments in science and technology have posed new issues. The ultimate challenge to philosophers, ethicists, theologians, and legal scholars is how to continue to apply the wisdom of the past to these new situations.

My initial interest in the golem was neither academic nor professional. The legend of the golem was but one of a myriad of Jewish legends I heard as a child that have always engaged me. In the 1970s I began work on what I hoped would be a comprehensive study of the life and thought of Judah Loew of Prague. Precisely because he is popularly identified as the legendary creator of the golem, I focused my attention on the nature of his thought and on his massive writings in theology, mysticism, ethics, and law. Believing him as I do to be the most comprehensive, important, and influential Jewish thinker in the long history of the Ashkenazic—that is, European Jewish—tradition, my chief aim was to describe the nature and importance of his life and thought apart from the legend that had served to obscure them—the legend of the golem, which was not even ascribed to him until centuries after his death. The main product of my efforts was a book entitled *Mystical Theology and Social Dissent: The Life and Thought of Judah Loew of Prague*, published in 1982. Yet, despite my efforts to deal with Judah Loew apart from the legend of the golem, many of my students, colleagues, and friends, encouraged me to deal with Loew and the legend.

The historical background of the golem legend in classical Jewish literature is amply dealt with in the seminal essay by the late Gershom Scholem of the Hebrew University of Jerusalem, who in the 1920s established the academic study of Jewish mysticism. Scholem's study "The Idea of the Golem" may be found in English translation in his *On the Kabbalah and Its Symbolism* (1965). By then Scholem had already compared the computer to the golem, which reinforced my intuition about the relevance of the golem legend to unfolding developments in science and technology. See Scholem's "The Golem of Prague and the Golem of Rehovot" in his 1971 volume *The Messianic Idea in Judaism*.

My conviction regarding the implications of the golem legend for dealing with the ethical problems engendered by the emergence of bioengineering was pushed a step farther by my teacher, colleague, and friend, the late Rabbi Seymour Siegel of the Jewish Theological Seminary, who served in the early 1980s as a member of the President's Commission for the Study of Ethical Problems in Medicine and Biomedical and Behavioral Research. That commission had been assigned by President Carter to deal with the bioethical implications of the new technology in the wake of the Supreme Court's landmark decision in *Diamond v. Chakrabarty*.

My interests were further stimulated and reinforced by the studies in the 1960s and 1970s of the rabbi and computer scientist

214

Azriel Rosenfeld. He was among the first to relate robotics and genetics to the golem. His essays were published in the journal of the American Orthodox rabbinate, *Tradition*: "Religion and the Robot" (1966), "Judaism and Gene Design" (1972), and "Human Identity: Halakhic Issues" (1977). More recently, attitudes and views in contemporary Jewish law and ethics on bioengineering and reproductive biotechnology have been summarized and discussed in Miryam Wahrman's *Brave New Judaism* (2002).

The spiritual dangers implicit in the Western philosophical and scientific tradition of human beings thinking of themselves as a type of machine were pointed out to me in the 1960s when I was a graduate student in philosophy at New York University. My teacher was the late William Barrett, who was instrumental in introducing existentialist thought in America. He dealt with these issues in his *Death of the Soul: From Descartes to the Computer* (1986) and *The Illusion of Technique* (1978).

In the early 1980s I began to investigate the golem legend in terms of its influence on modern and contemporary literature, art, and science, and its relevance for the ethical implications of newly emerging technologies. The more I investigated the matter, the more I realized that even though golems are usually described as being devoid of speech, they nonetheless have a lot to say about recent and expected advances in science and technology.

In writing this book I have both drawn and expanded upon my work on the golem legend for the past twenty or more years. This includes numerous papers delivered at conferences, interviews on radio and television, popular lectures and various publications. Among the latter are *The Golem Legend: Origins and Implications* (1984) and chapters in *In Partnership with God: Contemporary Jewish Law and Ethics* (1990) and *Jewish Ethics for the Twenty-First Century* (2000).

I first dealt with the ethical implications of the golem legend for developments in reproductive therapy in my 1995 essay "The Golem, Zevi Ashkenazi, and Reproductive Biotechnology," published in the journal *Judaism*. But with the cloning of Dolly the sheep I was motivated to relate the golem legend to the bioethical issue of human reproductive cloning when I was invited to speak in 1997 at the Capitol in Washington, at a conference sponsored by Senators Joseph Lieberman and Harry Coates on the ethical and legal implications of cloning technology. With the subsequent publication of the human genome, I was invited to comment at a variety of conferences and in a number of anthologies on the implications of genomics.

Classical Jewish sources on the golem legend that I refer to in this book are cited in detail in my earlier writings. Gershom Scholem's works should also be consulted. But the definitive study of the golem in classical Jewish sources is the magisterial work *Golem: Jewish Magical and Mystical Traditions on the Artificial Anthropoid* (English 1990, Hebrew 1996) by my friend and colleague Moshe Idel of the Hebrew University.

Many of the modern and contemporary works of literature that relate to the golem are noted above in Chapter Four. Two important studies on the influence of the golem legend on contemporary literature and culture are Arnold Goldsmith's *The Golem Remembered* (1981) and Emily Bilski's edited volume *Golem!: Danger, Deliverance, and Art* (1988), produced in conjunction with a remarkable exhibit on the golem at the Jewish Museum in New York. Gershon Winkler's *The Golem of Prague* (1980) contains in very readable form many of the modern versions of the golem legend found in the influential work by Yudel Rosenberg, *The Wonders of Rabbi Loew* (1909), and Chaim Bloch's *The Golem: Legends of the Golem of Prague* (1925). The academic studies of Hillel Kieval and of V. Sadek are also helpful. Kieval's study "Pursuing the Golem of Prague" appeared in the journal *Modern Judaism* in 1997 and Sadek's in the journal *Judaica Bohemiae* in 1987. Both speculate about how and why the golem legend came to be associated with the figure of Rabbi Loew.

There is no shortage of available information about emerging technologies or their implications for ethical and public-policy decision-making. Articles appear almost daily in newspapers and magazines. A wealth of information is available on a myriad of internet websites, some of which are cited in the recent books noted below.

A good place to begin is the 1982 landmark report of the President's Commission for the Study of Ethical Problems in Medicine and Biomedical and Behavioral Research, *Splicing Life*. Here the scientific nature and ethical problems of genetic engineering technology—though now slightly dated—are set down clearly and precisely.

The implications of the new technologies, especially in biology and medicine, for human nature and human life in the twenty-first century are considered in Francis Fukuyama, *Our Posthuman Future: Consequences of the Biotechnology Revolution* (2002); Leon R. Kass, *Life Liberty and the Defense of Dignity: The Challenge for Bioethics* (2002); Finn Bowring, *Science, Seeds and Cyborgs: Biotechnology and the Appropriation of Life* (2003); and Bill McKibben, *Enough: Staying Human in an Engineered Age* (2003). These works mostly concentrate on the dangers that biotechnology poses to human nature and integrity and to the social fabric. The most persistent and vocal critic of the biotech industry has been Jeremy Rifkin, particularly in his *The Biotech Century: Harnessing the Gene and Remaking the World* (1988).

Two accessible presentations of genomics are Matt Ridley, *Genome* (1999) and Nicholas Wade, *Life Script: How the Human Genome Discoveries Will Transform Medicine and Enhance Your Health* (2001). On the nature and implications of genetically modified food, see Bill Lambrecht, *Dinner at the Gene Café* (2001); Paul Lurquin, *High Tech Harvest* (2002); and Gregory Pence, *Designer Food* (2002). On genetic engineering and weapons of mass destruction, see, for example, the sobering account in Judith Miller, Stephen Engelberg, and William Broad, *Germs: Biological Weapons and America's Secret War* (2001). For a cutting-edge discussion of where genetic therapies may be headed, Gregory Stock's *Redesigning Humans: Our Inevitable Genetic Future* (2002) is both stimulating and unsettling.

In the 1960s the Nobel Prize Laureate Joshua Lederberg was among the first to discuss the implications of genetic engineering and cloning. Among the first philosophers to deal with these issues was Hans Jonas in his seminal essay "Bioengineering—A Preview," published in 1974 in his *Philosophical Essays*. Paul Ramsey, an eminent theologian and one of the founders of the field of bioethics, dealt with such issues in *Fabricated Man: The Ethics of Genetic Control* (1970). Each of these thinkers greeted developments in bioengineering with dismay and apprehension. Each both influenced and reinforced the views of Leon Kass, appointed by President George W. Bush as chairman of the President's Council on Bioethics. The report of the council was published in 2002 as *Human Cloning and Human Dignity*. For a variety of views on the nature and implications of stem-cell research and its potential applications and implications, see the edited volume by Suzanne Holland, Karen Lebacqz, and Laurie Zoloth, *The Embryonic and Stem Cell Debate: Science, Ethics and Public Policy* (2001). In essays included in *Public Policy and Social Issues: Jewish Sources and Perspectives* (2003), edited by Marshall Breger, and in the *2001 Proceedings of the Rabbinical Assembly*, I offer some of my views—informed by Jewish sources—on genetic engineering, cloning, genetic therapies, stem-cell research, and related issues.

An accessible introduction to the scientific development of cloning and to the ethical and public-policy issues it raises is Gina Kolata's *Clone: The Road to Dolly and the Path Ahead* (1998). Princeton professor Lee Silver's *Remaking Eden: Cloning and Beyond in a Brave New World* (1997) has become an often-used starting point for discussions of cloning and genetic technologies and their implications.

On the legal, ethical, and public-policy issues stimulated by developments in reproductive biotechnology, cloning, and genetics, I have been informed over the years by the groundbreaking work of the Chicago legal scholar Lori Andrews and by the pleasure of conversations with her. See, for example, her *The Clone Age: Adventures in the New World of Reproductive Technology* (1999); *Future Perfect: Confronting Decisions about Genetics* (2001); and, with Dorothy Nelkin, *Body Bazaar: The Market for Human Tissue in the Biotechnology Age* (2001).

A Note on Sources and a Personal Note

Robots (1978), edited by Harry Geduld and Ronald Gottesman, traces the history of the creation of machinelike creatures and their depiction in legend, literature, and film, including the golem legend, and provides a wealth of information, including many visual images. Geoff Simon, *Are Computers Alive?: Evolution and New Life Forms* (1983), clearly sets out the issues regarding "robotic life" and is still of value. More recent is Thomas Georges's *Digital Soul: Intelligent Machines and Human Values* (2003). The influential works of robotics and computer scientists—Rodney Books, *Flesh and Machines* (2002); Ray Kurzweil, *The Age of Spiritual Machines* (1999), and Hans Moravec, *Mind Children: The Future of Robot and Human Intelligence* (1988)—have elicited great anticipation and apprehension about developments in robotics, artificial intelligence, and computer science. Although now somewhat dated, Stephen Levy's, *Artificial Life: The Quest for a New Creation* (1982) provides a forthright and informative survey of the unfolding field of "artificial life."

Lori Andrews, Jeremy Rifkin, Bill Lambrecht, and others noted above have dealt with many of the implications of genetic engineering and biomedical technology for corporate ethics. In *Unequal Protection: The Rise of Corporate Dominance and the Theft of Human Rights* (2002), Thom Hartmann has traced the development of corporate abuses of power, the personalized status of corporations, and the dangers they portend now and in the future. On contemporary understandings of the corporation, a clear, precise, and informative analysis may be found in a 1992 article by Michael J. Phillips, "Corporate Moral Personhood and Three Conceptions of the Corporation," in *Business Ethics Quarterly*. On attitudes toward and views of the corporation in Jewish law, see Michael Broyde and Steven Resnicoff, "Jewish Law and Modern Business Structures: The Corporate Paradigm," in *Wayne Law Review*, Fall 1997.

New books and new technological developments will continue to appear as our "age of the golem" unfolds. What the Bible says about books may also become true of golems: of their making, there is no end.

Index

Abulafia, Abraham, 43

Achron, Joseph, 39

Adam, 7, 8, 16, 30, 49, 119; and asexual reproduction, 117, 118; and Eve, 117, 118; fire, invention of, 203; as golem, 115, 118

Adelphia Communications, 179

Adler, Alfred, 3

African Americans, 164

AI (film), 127, 137, 138

AIDS, 84

Amazing Adventures of Kavalier and Clay, The (Chabon), 37

American College of Medical Genetics, 87

American Fertility Society, 114

American Medical Association, 150; and code of ethics, 174; and patenting procedures, 87

Andrews, Lori, 87, 176

Animal rights activists: and genetic engineering, 78, 79; and pharmaceuticals, 78; and speciesism, 79

Anthrax, 83

"A Propos of the Golem" (Hollander), 39

Aquinas, Thomas, 10

Aronson, David, 42, 43

Art, 41; as golems, 37

Arthur Andersen, 180

Artificial intelligence (AI), 35, 46, 62, 144, 155

Index

Artificial Intelligence Laboratory, 126

Artificial Life (Levy), 199

Artificial life: and computer science, 64; implications of, 63–64; as new science, 63

Ashkenazi, Zevi, 16, 186, 188; and golem, legal status of, 15, 115, 116, 118

Asimov, Isaac, 211

Asner, Ed, 41

Astrology, 202

Atomic bomb, 82, 152

Auden, W. H., 205

Augustine, 48

Auschwitz, 167, 168, 169, 170, 172; and Auschwitz-Birkenau, 169; and Auschwitz-Buna, 169; and Auschwitz-Monowitz, 169

Australia, 98

Babylonia, 200

Banquo, 15

Barrett, William, 141, 142, 146

BASF, 170. *See also* I. G. Farben.

Batman Beyond, 194

Bayer, 170. *See also* I. G. Farben.

B. cepacia: commercial value of, 58; description of, 58; ecological value of, 58

Ben Gurion University (Israel), 133

Berkshire Hathaway, 181

Bible, 147; and biblical religion, 201, 202

Bilski, Emily, 37–38

Biochemical weapons, 82; and Richard Nixon, 82; and viruses, 82

Bioengineering, 46, 55, 59, 60, 63, 82, 214

Bionics, 135, 140, 155

Biotechnology, 46, 55, 59, 62, 79, 135; growth of, 80; and human cloning, 100; and patenting, 87–88

Black, Justice, 184

Blackstone, William, 156, 157, 171, 186, 189

Blade Runner (film), 86

Blair, Tony, 100

Blavatsky, Madame, 39

Block, Hayyim, 24

BMW, 167

Body Bazaar (Andrews and Nelkin), 176

Bohemia, 22, 27, 40

Boltanski, Christian, 43

Book of Splendor, The (Sherwood), 39

Borges, Jorge Luis, 39

Borkin, Joseph, 168

Brandeis, Louis, 158, 159, 178, 182, 191

Brandeis University, 45

Brave New World (Huxley), 60, 95

Brooks, Rodney, 126, 127, 134–135, 138
Brown, John, 96
Brown, Lesley, 96
Brown, Louise, 96, 97
Buffett, Warren, 181
Burger, Warren, 58, 59
Bush, George W., 100, 218; and stem cell research, 81, 101
Butler, Samuel, 127, 128
Byron, Lord, 197

Caplan, Arthur, 90
Carson, Rachel, 68
Carter, Jimmy, 192
Casken, John, 40
Casper, the Friendly Ghost, 44
Catechism of the Catholic Church, The: and reproductive biotechnology, 117
Cayman Islands, 177
Cepak brothers, 126, 139
Chabon, Michael, 37, 43
Chakrabarty, Ananda, 57, 58, 64
Chargoff, Erwin, 172
Charles, Prince, 73
Chernobyl, 69
China, 70
Christie, Julie, 137
Civil War, 160, 161, 164
Clarke, Arthur C., 57
Cleveland, Grover, 165
Clinton, Bill, 86, 100

Cloning, 46, 60, 81, 95, 97, 101; and aging, 104; benefits of, 109; criminalization of, 108; and Dolly the sheep, 72, 99, 100, 105, 110, 215–216; and embryos, 111, 112; ethical implications of, 111, 215–216; and Frankenstein, 120; of frogs, 98; of horses, 196; of human beings, 100, 101, 102, 105, 106, 107, 108, 118, 119; and identical twins, 119; and in vitro fertilization (IVF), 99; as immoral, charges of, 99; Jewish bioethicists, attitude toward, 105, 106, 107, 113; medical implications of, 101; moral implications of, 101; opponents of, 116, 119; procedure, description of, 98, 99; and public opinion, 72, 100, 102; and stem cells, 101, 102, 103, 104, 108, 109, 111, 113; theologians, attitude toward, 105; and therapeutic human cloning, 102, 105, 108, 110
Collins, Harry, 35–36, 143
Colossus: The Forbin Project (film), 139
Commentaries on the Laws of England (Blackstone), 156, 189
Computers, 131, 138, 141, 150; and artificial life, 64, 134; and

Computers (*cont.*)
biology, 133, 134; and chess, 137, 145; and computer science, 35, 62, 131, 135, 150, 155; and corporations, 134; and DNA chips, 133; and human evolution, 136

Computing Machinery and Intelligence (Turing), 132

Concentration camps, 167, 168

Constant Gardener, The (Le Carré), 172

Copernican Revolution, 50

Corporations: abuses of, 164, 177, 182; and accounting firms, 179; and American law, 181, 182, 183, 188; amorality of, 189; as artificial persons, 156, 157, 161, 162, 163, 184, 185, 189; benefits of, 158; and boards of directors, 178, 179; and British common law, 185, 186; and Civil War, 160; control of, 181, 182, 185, 189, 191; and Corporate Responsibility Act, 177–178; and corruption, 161; crimes of, 177, 189, 190, 191; dangers of, 158; and developing countries, 80; enemies, trading with, 170, 171; and ethics, 46; farmers, affect on, 79, 80; food industry, 79; foreign policy, manipulation of, 80; and Fourteenth Amendment, 163, 164, 165, 184, 187; and fraud, 177, 180; in Germany, 168; and golem legend, 182; and government, influence on, 185; and governments, sovereignty of, 176; greed of, 171; history of, 160; and the Holocaust, 168; human rights, violations of, 177; and Jewish law, 181, 186, 187, 188; legal status of, 182, 185, 186; and lobbyists, 180; and moral responsibility, 167, 178; and multinationals, 80, 176, 177; as natural persons, 183, 184, 185, 191; and pharmaceutical industry, 172, 175; politics, influence on, 180; power of, 157, 159, 160, 165, 171; punishment of, 178, 180, 181, 190, 191; and railway industry, 161, 162, 163; reform, call for, 177, 178; and Theodore Roosevelt, 166; scandals of, 72, 158, 176, 179, 180, 181; and stockholders, 178, 187, 188; and Supreme Court, 164, 165, 180; and tax evasion, 177; Trading with the Enemies Act, 170; during World War II, 170. *See also* Health-care industry; Individual corporations.

"Cosmos" (television series), 63

County of Santa Clara v. The Southern Pacific Railroad, 162, 163, 164, 183, 184, 186; and Grover Cleveland, 165

Creation: and God, relationship between, 53, 54, 55

Creativity: and control, loss of, 38; danger of, 17; and golem, legend of, 37, 38, 43; and Jewish tradition, 121

Crichton, Michael, 82, 151, 152, 153

Crime and Punishment of I. G. Farben, The (Borkin), 168

Crops: and nutrition, 68, 69; protection of, 68; and weather conditions, 68

Cybernetics, 44, 132

Cyborgs, 135, 137, 140

D'Albert, Eugene, 39

Daimler-Benz, 167

Dan, Joseph, 25

Darwin, Charles, 51

Davenport, Charles, 91

Davidson, Avram, 44

Davis v. Davis, 115

Davis, J.C. Bancroft, 163

Death of the Soul: From Descartes to the Computer, The (Barrett), 146

Delmas, Delphin, 186

Demon Seed (film), 137

Der Goylem (Leivick), 39, 42

Descartes, René, 128, 129, 145

Dexler, Eric, 154

Diamond v. Chakrabarty, 57, 80, 174, 192, 214

Diamond, Sidney, 57, 58

Dinesen, Isak, 130

DNA, 48, 50, 57, 64, 72, 96, 202; and chips, 133; and fingerprinting, 56; and genetic engineering, 52, 55, 56; and genetically modified food, 67; and genetic testing, 56; and weapons of mass destruction, 82

Doyle, Arthur Conan, 24

Drugs: and animals, use of, 78; commercialization of, 175, 176; manufacture of, 78; and patenting, 175; prices of, 175, 176; and taxpayers, 175

Dungeons and Dragons, 44

DuPont, 170

Dybbuk, 5

Dyson, Freeman, 83, 133

Edwards, Dr. Robert, 96, 97, 99

Egypt, 200

Eisenhower, Dwight, 166, 167, 169, 170, 171

Elijah of Helm, 16, 17, 24, 33, 45, 148, 188

Eliot, T. S., 206

Elkins, William F., 160
Embryos: and asexual
reproduction, 118; and ethics,
114; as golem, 114; and Jewish
law, 113, 114; and personhood,
114; as property, 114; status of,
111, 112, 115, 116
Emden, Jacob, 15, 16, 116, 129,
148
Engel, George, 130
*Enough: Staying Human in an
Engineered Age* (McKibben),
127
Enron, 177, 179
Epic of Gilgamesh, 147
Erin Brockovich (book and film),
181
The Ethics of the Fathers, 207
Eugenics, 90; advocates of, in
America, 91
European Jews, 27, 28
European Union, 76
Eve: and asexual reproduction,
117, 118
Eve, Nami, 39
Evolution: theory of, 51
Extinctionists, 147

Family Orchard, The (Eve), 39
Farben, I. G., 167, 168, 171;
chemical weapons, production
of, 169–170; and DuPont, 170;
and German war effort, 169,

170; and Standard Oil, 170; war
crimes, 170
Farmer, James Doyne, 63, 64, 199
Fate, 201; and astrology, 202
Faust: legend of, 10, 147
Field, Justice, 190
Fifth Amendment, 184
First Amendment, 180, 185
*First National Bank of Boston v.
Bellotti*, 180, 183, 185
*Flesh and Machines: How Robots
Will Change Us* (Brooks), 126,
135
Ford Motor Company, 170, 171
Foresight Institute, 154
Fourteenth Amendment, 162, 163,
164, 165, 183; and corporations,
184, 186, 187
France, 171
Frankenstein (Shelley), 10, 11,
120, 159, 194, 205; and golem
legend, 196, 200; and Greek
mythology, 196, 201; hubris in,
197; lessons of, 199
Frankenstein, Victor, 19, 24, 33,
40, 46, 120, 156, 193, 194, 209,
212; divine powers,
presumption of, 197; and
genetically modified food, 79;
as golem, 207; hubris of, 197,
198, 199; and Judah Loew, 194,
195, 196, 198
Frankfurt (Germany), 40

Franklin, Benjamin, 72
Freundel, Barry, 93
Friedman, Milton, 189
Frischman, David, 39
Future Shock (Toffler), 135

"Galactic Golem," 44, 194
Galileo, 48
Gattaca (film), 86
Gaylin, Willard, 99
Gelsinger, Jesse, 88
Genetic engineering, 35, 52, 61,
    62, 93, 119; animals, suffering
    of, 78, 79; benefits of, 59, 70,
    84, 85; and Jimmy Carter,
    192–193; and cattle, 70; and
    Ananda Chakrabarty, 57, 192;
    and chickens, 70; and cloning,
    60; consequences, long-term,
    85; and corporate greed, 79;
    and crop protection, 68; and
    cross-breeding, 77; dangers of,
    59, 73, 82; development of, 55;
    and disease, 69, 79, 84, 85, 88;
    divine prerogative, usurpation
    of, claims of, 73, 74; and
    environment, 78; and eugenics,
    90; Europeans, suspicion
    toward, 77; and fatalism, 202;
    and fish farming, 69, 70; and
    food, 66, 67, 68, 70, 71; and
    frankenfood, 79; genetic-
    enhancement therapy, 89; and

gene therapy, 84, 85, 88; and
    growth hormones, 89, 90; and
    human beings, 83, 84, 86; and
    human genome, 84;
    interspecies genetic transfers,
    60; and natural foods, 73; and
    nutrition, 68, 69; opponents of,
    73; and papaya industry, 68;
    and patent law, 80, 81, 87;
    pharmaceuticals, manufacture
    of, 70; and plant biotechnology,
    69; and pollution, 69; and
    public policy, 59, 70; purposes
    of, 56; reaction toward, 56;
    risks of, 77, 78; and science
    fiction, 85; and speciesism, 79;
    and trans-species organ
    transfers, 60
Genetic screening, 86, 87; Jewish
    tradition, attitude toward, 88;
    and Tay-Sachs disease, 87
Genetically modified food
    (GMF), 66, 67; animals,
    suffering of, 78; Prince
    Charles, opposition to, 73; and
    corporations, 72; debate over,
    70; Europeans, aversion to, 73,
    76, 77; and famine, 77; and
    FlvrSavr tomato, 71; labeling
    of, 71, 72; and malnutrition, 77;
    monitoring of, 75; and natural
    food, 74, 75, 76; safety of, 74;
    as symbol, of American power,

Index

Genetically modified food (*cont.*)
76; and U.S.-European
relations, 72–73

Genetically Organized Lifelike
Electro Mechanics (GOLEM),
45

Geomorphically Orogenic
Landscape Evolution Model
(GOLEM), 45

Gerke, Achim, 76

Germany, 27; corporations in,
167, 168

Gershuni, Judah, 97

Gersonides, 48

God and Golem, Inc. (Wiener),
44, 125, 126

Goldsmith, Arnold, 216

Golem: and Adam, 7, 8, 115; and
art, 37, 42, 43; as artificial
man, 10, 41; and artificial
intelligence (AI), 35, 46; and
asexual reproduction, 117, 118;
and the Bible, 7, 8, 11; and
bioengineering, 46; and
biotechnology, 46, 64, 65; and
Hayyim Block, 24; and blood
libel, 24; as blueprint, 115; and
Michael Chabon, 37;
characteristics of, 8, 15; and
childhood, 3; and children's
books, 38, 42; and children's
games, 44; and cloning, 46,
112; and comic books, 44;

communion with God, 45; and
computer science, 35, 45, 214;
control of, 182; and
corporations, 157, 158, 182; and
corporate ethics, 46; and the
creation, 7, 8, 12; and
creativity, 37, 38, 43, 210; and
dance, 39; as dangerous, 14,
17, 19, 24, 32, 44, 46;
deactivation of, 31, 32; as
defender, 31; depiction of, 19,
37, 38, 42; divinity of, 197;
DNA, 49; durability of, 26, 37;
and embryos, 112, 114, 115; as
evil, 44; and fascism, 40; as
female, 124; in film, 40;
as Frankenstein, Jewish
facsimile of, 19, 46, 194; and
Frankenstein story, differences
between, 196, 199, 200; and
genetic engineering, 35; and
The Golem (novel), 6; as first
human being, 8, 9, 10, 11;
in Hebrew, 115; historical
background of, 214; human
status, discussion of, 15, 16,
118; and Solomon ibn Gabirol,
123, 124; and in vitro
fertilization, 97; and Israel, 41;
in Jewish literature, 15, 39, 114,
115; legal status of, 15, 16, 115,
116, 129; legend of, 3, 5, 6, 12,
14, 16, 17, 20, 22, 23, 24, 25, 45,

228

46, 49, 120, 159, 193, 200; as liberator, 40; and Judah Loew, 12, 17, 18, 19, 22, 23, 26, 27, 28, 29, 30, 31, 33, 92, 119, 193–194; and machines, 124, 127, 143, 155; and magical activities, 10; meaning of, 8, 9; and mechanized society, 42; and medieval Jewish mystics, 12, 13, 14; as metaphor, 34, 35, 42, 46; modern version of, 19, 24, 32, 35; and music, 39; as myth, enduring, 25; and nation-states, 41; popularity of, 25; of Prague, 17, 18, 19, 22, 23, 26, 27, 29, 30, 31, 32, 37, 45, 126, 148, 149, 194; name of, 24; and novelists, 37, 39; and opera, 39, 40; and plays, 42; and poems, 39; and pogroms, 31; in pop culture, 38, 39; as power personified, 31; as protector, 14, 19, 24, 41, 45; prototype of, 19, 22; purposes of, described, 14, 17, 32, 149; and Rava, 13, 14, 148; replication of, 148, 153; relationship, between divine and human, 42–43; and robotics, 35, 45, 46; and Yudel Rosenberg, 23, 24, 25; as savior, 19, 24, 41; and science, 36, 42, 44, 143; and science fiction, 44, 139; as servant, 17, 19, 24, 30, 45; and sexuality, 149; and Isaac Bashevis Singer, 37; and "Sorcerer's Apprentice," 24; and speech, 13, 14, 47, 48; as spy, 31; as symbol, 42; and Talmud, 207; and technology, 36, 42, 44; on television, 40, 41; and universal problems, 25; as warrior, 19; Western culture, influence on, 34; and Elie Wiesel, 38; and wisdom, 206, 207, 210; and Zionism, 41

Golem and Tselem (artwork), 43

Golem (artwork), 43

Golem at Large: What You Should Know About Technology, The (Collins and Pinch), 35–36

Golem: How He Came into the World, The (film), 40

Golem in L.A., The (film), 41

Golem, Le (film), 40, 42

"Golem Project, The" 45

Golem Records, 43

"Golem Shuffle, The" (Gottshall), 40

"Golem Suite, The" (Achron), 39

Golem, The (Meyrink), 6, 39

"Golem, The" (short story), 44

Golem: What You Should Know About Science, The (Collins and Pinch), 35–36

Golem, Yossele, 24, 30, 31, 32

Index

Golems of Gotham, The
  (Rosenbaum), 39
Gonen, Jay, 41
Gottshall, Dan, 40
Great Britain: Mad Cow disease
  in, 71

Hadassah Medical School
  (Jerusalem), 105
Hale v. Henkel, 182
Ha-Meiri, Menahem, 203
Hamill, Pete, 40
Hanina, Rabbi, 66, 80
Harding, Warren G., 166
Harvard University, 56
Hawaii, 68
Hawking, Stephen, 44, 137
Health-care industry, 171, 172;
  and artificial insemination, 96;
  and artificial wombs, 95; and
  donor insemination, 95, 96; egg
  donations, 95; and infertility,
  95; and patents, 175; and in
  vitro fertilization (IVF), 95, 96,
  97
Hebrew: and alphabet, 47, 48;
  and creativity, 13, 48; and
  wisdom, 206
Hebrew University of Jerusalem,
  126, 216
Henkel v. Henkel, 165
Hitler, Adolf, 169, 170
Hobbes, Thomas, 41

Hollander, John, 39
Holmes, Oliver Wendell, 91, 92
Holmes, Sherlock, 24
Holocaust, 4, 167; and
  corporations, in Germany, 168
Holy Roman Empire, 27–28
Hudock, George, 119
Human beings: and animals, 129,
  130, 203; and biopiracy, 174;
  and creation, 12, 13; character,
  reflection of, 13; characteristics
  of, 137; and cyborgs, 135; and
  René Descartes, 129; divinity,
  confusion with, 209, 210;
  economic exploitation of, 174,
  175; extinction of, 153; and
  fatalism, 202; and genetic
  screening, 86, 87; and
  genocide, 147; as God's partner,
  73; and golems, 9, 10, 11, 13,
  15; and human body,
  demeaning of, 145, 146; and
  intelligence, 144; and Julien
  Offray de la Mettrie, 129; and
  machines, 124, 125, 127, 128,
  130, 132, 135, 136, 137, 140, 141,
  142, 143, 144, 145, 146, 154,
  155; and the mind, 145; and
  mortality, 146, 147; and nature,
  54, 202, 203; and patenting, 81;
  and the soul, 144, 145; stature,
  deflation of, 51; superiority of,
  51

Human genome, 48, 86, 134; and
DNA, 48, 50; ethics of, 208;
and genetic engineering, 84;
and gene therapy, 49, 85; and
golem legend, 49; and
patenting, 81; sequencing of,
50; and speciesism, 51, 52
Human Genome Project, 175
Human reproduction: and Adam
and Eve, 117; artificial
insemination, 116–117; and
asexual reproduction, 117, 120;
and Catholicism, 117; and
cloning, 117; and in vitro
fertilization, 117; and Jewish
law, 117
*Human Use of Human Beings,
The* (Wiener), 196
Huxley, Aldous, 60, 95
Hydrogen bomb, 82, 148

IBM, 170
Ibn Gabirol, Solomon:
background of, 123–124; and
golem legend, 123, 124, 130
Infertility, 94; and health-care
industry, 95; and intra-
cytoplasmic sperm injection
(ICSI), 97–98; and in vitro
fertilization (IVF), 95, 96, 97,
98; and reproductive cloning, 98
Infertility in Women (Kleegman
and Kaufman), 97

Ingersoll, Robert, 130
Iraq: war against, 72, 73; and
European opposition to, 76
Israel, 200

Jesus: and asexual reproduction,
117
Jewish ethics: and cloning, 105
Jewish law: and asexual
reproduction, 118; and cloning,
107; and conception, 113; and
embryos, 112, 113
Jewish mystics: and art, 37; and
artificial life, 64; and creation,
48; and Hebrew alphabet, 47
Jewish tradition: and artificial
life, 120; and cloning, 105; and
creativity, 11, 121; and life,
creation of, 25; mystical
tradition of, 11, 12, 13, 14, 25,
28, 37; and *Sefer Yetzirah* (The
Book of Creation), 13
Jews: and blood libel, charge of,
28, 31; expulsion of, 21, 27; and
Judah Loew, 28; Ottoman
empire, alliance with, 27; and
pogroms, 31; in Prague, 21, 31;
in Spain, 27; in Vienna, 27
Joy, Bill, 155; and technology,
danger of, 150, 151, 152, 153
Judaism, 121; and moral choice,
202; nature, human
intervention of, 204, 205; and

Judaism (*cont.*)
technology, 205; and wisdom,
205; and the world,
improvement of, 83

Kabbalah, 20, 28, 29
Kahn, Toby, 43
Kant, Immanuel, 142
Kasparov, Garry, 137, 145
Kass, Leon, 100, 101, 104, 117
Katz, Isaac, 28
Kaufman, Sherwin, 97
Kepler, Johannes, 128
Kessler, Dr. Jack, 109
Kieval, Hillel, 216
King Arthur, 3, 5
King David, 30
Kleegman, Sophia, 97
"Knight Rider" (television series),
144
Knights of the Round Table, 3
Krupp, 167
Kubrick, Stanley, 137
Kurzweil, Ray, 127, 134, 135, 136,
143, 145, 147, 150, 152, 153; and
fatalism, 202

La Mettrie, Julien Offray de, 129
Language, power of, 13
Le Carré, John, 172
Lederberg, Joshua, 99
Lehrer, Tom, 207
Leiner, Gershon Hanokh, 15, 16,
118

Leivick, Halper, 39, 42
Leviathan, The (Hobbes), 41
Levy, Steven, 199
Life: and artificial forms, 57, 63;
and cloning, 112; and
conception, 111, 112; creation
of, 10, 12; and Christian
tradition, 10; definition of, 63;
and the devil, 10; and Faust
legend, 10; genetic engineering,
11; and genome, 48; and
golems, creation of, 20; and
Jewish mystical tradition, 11,
25; Jewish view of, 11; and
patents, 58, 59; speech, power
of, 13; and Talmudic tradition,
10. *See also Diamond v.
Chakrabarty.*
Liggett v. Lee, 159
Lincoln, Abraham, 160, 165, 171;
corporations, corruption in,
161
Lochner v. New York, 164
Loew, Hava, 21, 22, 149
Loew, Judah, 11–12, 17, 26, 41,
42, 49, 53, 57, 119, 121, 149,
156, 158, 191, 193, 199, 202,
209, 210–211, 212, 213; death
of, 18, 22, 33; fame of, 19; and
Victor Frankenstein, 194, 195,
198; and golem legend, modern
version of, 19; and golem of
Prague, 18, 19, 22, 23, 27, 28,

29, 30, 31, 32, 33, 92, 194, 195; and gravesite, 22; and John Hollander, 39; and Jewish mystical tradition, 28; legends of, 21, 22; and Miriam (daughter of), 149; as national hero, 22; and nature, 204; in Prague, 20, 22; and Yudel Rosenberg, 23, 24; and Rudolph II, 20, 21; sculpture of, 22; wisdom of, 207; work of, 18

Los Alamos (New Mexico), 63

Luther, Martin: and Jews, expulsion of, 27

Macbeth (Shakespeare), 15

McCarthy, John, 144

Machine Nature: The Coming Age of Bio-Inspired Computing (Sipper), 133

Machines: and American medicine, 130; and computers, 125; as constant companions, 124, 125; and games theory, 148; and golems, 124, 126, 143, 148, 155; and humans, 124, 125, 127, 128, 130, 132, 135, 136, 137, 140, 141, 142, 143, 145, 146; and intelligence, 144; nature, duality of, 124; replication of, 148, 149, 150, 151, 152, 153, 154; and robots, 126, 127;

spirituality, affects on, 154; and war games, 148

McKibben, Bill, 127

Mad Cow disease, 73, 74; in Great Britain 71

Magnus, Albertus, 10

Maimonides, 202, 207

Man, A Machine (La Mettrie), 129

Massachusetts Institute of Technology (MIT), 56, 126, 138

Matrix, The (film), 139

Memories, 4, 5; and childhood, 3

Menahem ha-Meiri, Rabbi, 119

Mendy and the Golem (comic book), 44

Merlin, 5

Metamorphoses (Ovid), 197

Meyrink, Gustav, 6, 39, 43

Michelangelo, 30

Mind Children: The Future of Robot and Human Intelligence (Moravec), 147

Molecular Biology of the Cell, The (textbook), 133

Moore, John, 172, 173

Moore v. Regents of the University of California, 172, 173, 174

Moravec, Hans, 127, 143, 145, 147, 152, 153, 155; and fatalism, 202

Mosk, Justice, 174

Mulisch, Harry, 39, 49

Myths: purpose of, 25

Index

Nanotechnology, 152, 153, 155; description of, 150, 151; and nanorobots, 151, 154; and warfare, 151

Nature, 62, 75, 202, 203; and the Bible, 201, 202; and divinity, 12, 200, 201; and human intervention, 203, 204; and Nazism, 75, 76; and Scripture, 75

Nazism, 168, 169; and American eugenics movement, 91, 92; and corporations, 167, 170; eugenics policies in, 76, 90; and genocide, 92; natural selection, theory of, 75–76; nature worship, support of, 75; and racial science, 76

Nelkin, Dorothy, 176

New York Central and Hudson River Railroad Company v. United States, 189

Nifla'ot Maharal, The Wonders of Rabbi Loew (Rosenberg), 25; as forgery, 23

Nimrod, 209

9/11 attack, 81; and American economy, affect on, 72

Nixon, Richard, 82

Nuclear weapons, 152

Origins of Life, The (Dyson), 133

Oshaya, Rabbi, 66, 80

Out of Africa (Dinesen), 130

Ovid, 197

Ozick, Cynthia, 39, 94, 114, 149

Pacific Gas and Electric, 181

Paul v. Virginia, 164

People v. North River Sugar Refining Corporation, 188

Peretz, Y. L., 39

Physicalism, 130

Pincas, Abraham, 43

Pinch, Trevor, 36, 143

Plant biotechnology, 69

Plato, 145; and wisdom, 207

Pokemon, 44

Poland, 4

Polanyi, Michael, 141

Prague, 19, 20; and golem, 22, 33, 37, 148, 149; Jews in, 21, 22; and Old Jewish Cemetery, 22, 33

Presidential Commission on Cloning, 93

President's Commission for the Study of Ethical Problems in Medicine and Biomedical and Behavioral Research, 119, 192, 193

President's Council on Bioethics, 100, 101; and embryos, 112

Prey (Crichton), 151, 153

Procedure, The (Mulisch), 39, 49

Prometheus, 196, 197, 198, 203, 209

"Prometheus Unbound" (Byron), 197

Promethia, 196

Psychohistory of Zionism, A (Gonen), 41

Puttermesser Papers, The (Ozick), 39, 94, 114, 149

"Puttermesser and Xanthippe" (Ozick), 39

Ramsey, Paul, 99

Rathenau, Walter, 42

Rava, 10, 12, 15, 16, 18, 41, 66, 129, 144, 148; and golem, creation of, 13, 14, 47, 204

"Recipe for Life, The" (Chabon), 37

Rehnquist, William, 183, 184

Remaking Eden (Silver), 84

Reproductive cloning, 100, 101; arguments against, 108; ban on, 102; and Jewish religious tradition, 105, 106, 107

Richler, Mordecai, 23

Robin Hood, 4, 5

Robotics, 35, 46, 52, 62, 135, 150, 155, 157; and Isaac Asimov, 211; danger of, 152; and golem, 215; humans, relationship between, 126, 127; and nanorobots, 136, 152

Rohleder, Herman, 96

Rollerball (film), 176

Roosevelt, Franklin, 166

Roosevelt, Theodore, 91, 166, 171, 180, 183

Rosenbaum, Thane, 39

Rosenberg, Yudel, 23, 25; and golem, description of, 24; and Judah Loew, 24

Rosenzweig, Franz, 146

Rosner, Fred, 88, 89

Rothberg, Abraham, 149

Royal Library of Metz, 23

Rudolph II, 20, 28; and Judah Loew, 20, 21

R.U.R. (Cepak brothers), 126, 139, 140, 143, 148

Sagan, Carl, 51, 63

Salk, Jonas, 174, 206

Saloun, Ladislav, 22

Sanger, Margaret, 91

Sasson, Rabbi, 28

Scholem, Gershom, 45, 126

Science, 134, 142–143, 205; as golems, 36, 42; and golem legend, 44, 143, 215; and morality, 54; and nature, 52, 53, 54; and wisdom, 206

Science fiction, 137, 138; and golem legend, 139

Scotland, 99

Sefer Yetzirah (The Book of Creation), 13, 20

Shakespeare, William, 15, 204

Index

Shefatiya, Amitai ben, 115
Shelley, Mary, 10, 120, 194, 196, 197, 207
Sherman Antitrust Act, 166
Sherwood, Frances, 39
Siemens, 167
Silver, Lee, 84, 86
Singer, Isaac Bashevis, 37, 38, 42, 43
Singer, Peter, 51
Sipper, Moshe, 133
Snow in August (Hamill), 40, 41
Snow White, 4, 5
"Sorcerer's Apprentice," 24
Spain, 27
Speciesism, 51, 52
Spielberg, Steven, 137
Splicing Life (report), 193
Standard Oil, 170, 171
"Star Trek" (television series), 44, 127, 137, 138
Star Wars (film), 137
Steinberg, Dr. Abraham, 105
Steinberg, Saul, 43
Stem cell therapy, 109; and disease, 101, 102, 103, 104; and embryos, 110, 111, 114; and in vitro fertilization (IVF), 101; morality of, 110; and spinal cord injuries, 110
Steptoe, Dr. Patrick, 96, 99
Stock, Gregory, 202

Stories, 3, 4; as family heirlooms, 5; and the Holocaust, 4; and memories, 5; and Poland, 4
Stormer, Horst, 150
Sud, Ira, 33
Sun Microsystems, 150
Superman, 44, 194
Supreme Court, 162, 163, 192; and corporate rights, 164, 165, 180, 184, 185, 189, 190
Sword of the Golem, The (Rothberg), 149

Technology, 205; divinity, confusion with, 209; as golems, 35, 42, 44; Judaism, attitude toward, 83; and morality, 54, 55, 208, 209; nature, improvement upon, 84; theologians, attitude toward, 54, 55; and wisdom, 206
Tennessee, 115
Terminator 3: The Rise of the Machines (film), 139
Test Tube Babies (Rohleder), 96
Thirteenth Amendment, 81
"Three Laws of Robotics" (Asimov), 211
Toffler, Alvin, 135
Tower of Babel: Jewish interpretation of, 208, 209; and Nimrod, 209

Tudor, Fredrick, 74
Turing, Alan, 131, 134, 140–141;
  and Turing Test, 132, 138
2001: A Space Odyssey (Clarke),
  57
2001: A Space Odyssey (film), 137
Tyco, 179

UCLA Medical Center, 172, 173
United Nations, 70; and human
  cloning, 100
U.S. v. White, 184

Vienna, 27
Von Braun, Wernher, 207
Von Neumann, John, 148

Waite, Morrison, 162, 163
Washington, George, 4
Watson, James, 96, 202
Weapons of mass destruction, 82,
  83; and anthrax, 83
Wegener, Paul, 40, 149
Weizenbaum, Joseph, 138

Weizmann Institute (Rehovot,
  Israel), 45, 126
Western philosophy: and human
  body, disparaging of, 145, 146;
  and mortality, 146
Wiener, Norbert, 44, 125, 196;
  and cybernetics, 132
Wiesel, Elie, 38, 42
Wilbur, Richard, 210
Winter's Tale, A (Shakespeare),
  204
Wired (journal), 150
Wisdom, 205, 206, 210, 211, 213
Wisniewski, David, 42
Wolfram, Stephan, 131
World War II, 76, 131, 167, 170

"X-Files, The" (television show), 41

Zambia, 70
Zeira, Rabbi, 10; and golem,
  destruction of, 13, 14, 15, 129
Zeus, 196
Zevi, Hakham. *See* Ashkenazi,
  Zevi.

A NOTE ON THE AUTHOR

Byron L. Sherwin is Distinguished Service Professor of Jewish Philosophy and Mysticism at the Spertus Institute of Jewish Studies in Chicago. Born in New York City, he received degrees from Columbia University, New York University, and the University of Chicago, and was ordained a rabbi by the Jewish Theological Seminary. He is an internationally recognized authority on Jewish theology, ethics, and mystical traditions, and the prize-winning author of twenty-three books, including *Crafting the Soul*, *Why Be Good?*, and *Jewish Ethics for the Twenty-First Century*. He is married with a son and lives in Chicago.